NCU – 1

FAMILY NAMES IN CURRENT USE FOR VASCULAR PLANTS, BRYOPHYTES, AND FUNGI

Regnum Vegetabile

a series of publications for the use of plant taxonomists published under the auspices of the International Association for Plant Taxonomy edited by Werner Greuter

Volume 126

ISSN 0080-0694

NCU – 1

Family Names in Current Use for Vascular Plants, Bryophytes, and Fungi

Edited by

Werner Greuter

on behalf of the
Special Committee on Names in Current Use

compiled by

**R. D. Hoogland, J. L. Reveal,
M. J. Crosby, R. Grolle, G. Zijlstra**
and
J. C. David

1993
Published for the
International Association for Plant Taxonomy
by
Koeltz Scientific Books
D-6240 Königstein, Germany

ISBN 3-87429-346-7 (Germany)
ISBN 1-878762-42-7 (U.S.A.)

INTRODUCTION

By Werner Greuter

This booklet includes lists of family names found to be currently used for organisms treated as plants under the *International code of botanical nomenclature*. Fossil plants and the groups assigned to the algae are not covered, but all non-algal extant plants and fungi are. Criteria for defining "current use" may have slightly varied (see the introductions to the individual lists), but no currently used name has been listed unless it was found to have been validly published.

For all groups covered except for the Musci the nomenclatural starting-point date is 1 May 1753. However, for practical purposes family nomenclature has long been considered to start with Jussieu's *Genera plantarum* (Jul-Aug 1789). This tradition has been adhered to for the purpose of the present lists, and for very good reasons. No cut-off date has been defined, and even some names that were validly published early in 1993 have been included. However, the literature coverage for the very recent past has likely been somewhat incomplete.

The actual and potential status of NCU lists

As they are presented, the family NCU lists are no more and no less than thoroughly compiled and checked inventories of names. They will I am sure be welcomed and used by many, since the listed names can be used with confidence as being validly published and legitimate from the date and with the authorship quoted, and since citation has been made to conform with recently published, internationally recognized standards. It is the wish of the editor and authors that they in turn be used and recognized as a standard for family nomenclature in the groups covered. As such, they can save time and labour of plant taxonomists and lead to greater uniformity and consistency in botanical literature.

The lists include a great number of names that few will want to adopt and most will hold to be synonyms. This is one point that must be made well understood: by refusing to exert taxonomic censorship the authors of NCU lists will inevitably appear to back an extreme splitter's view. In fact, no one's view is being backed. The lists just offer the raw material from which one may safely select the names one needs to designate one's families, however defined. Just check which of the genera listed by name in the "Type" field belong in your family, then select the oldest of the corresponding family names. If you draw a complete blank you will have to look for an appropriate unlisted name or, if none exists, validate a new family name on your own. But such a case will, we hope, only occur quite exceptionally.

There is an obvious danger in publishing such lists, however. Some may not use them for the purpose for which they are designed, but will invest their skill and efforts in proving them to be wrong. They will doubtless succeed in many cases – but to whose benefit? Family nomenclature is clouded by an even larger number of uncertainties than nomenclature at the lower ranks. Some of the grey areas are mentioned in more detail below. Let me mention here one such aspect in particular. When defining a family name, the *Code* writes: Names intended as names of families, but published with their rank designated by one of the terms "order" or "natural order" instead of "family", are treated as having been published as names of families. This looks like a case of circular reasoning, and not at all like a limpid rule. It is not surprising that it is often hard to apply such a paragraph objectively!

In my opinion there is a strong case for banning the uncertainty and abolishing the reward for futile nomenclatural digging exercises. The proposals made to the XV International Botanical Congress, to grant protected nomenclatural status to listed names in current use, would provide for the legal options that are needed, and under such options the present lists would be prime candidates for approval and consequent stabilization (of nomenclature, mind you, not taxonomy). Since the principles and effects of nomenclatural protection are not yet familiar to all, I will again mention them below. The full proposals for NCU protection have been published in *Taxon* (40: 669-677. 1991).

Protection versus conservation

Conservation is of course to override the effects of protection as proposed, being the stronger of the two measures. By conservation, it will be possible to give preference to the younger of two protected names when this is desirable. However, the options for conservation are less complete than those envisaged for protection. There are, in particular, two things that conservation cannot effect but protection could. First, when it is found (or at least believed) that a name is not validly published in the place from which it is conserved, then conservation fails, since it can make illegitimate names legitimate but not render invalid names valid. Second, when it is found that a conserved name has been validly published in an earlier place than the one cited, the conservation entry must be corrected accordingly, with consequent change of priority and often authorship: while a name can be conserved against earlier homonyms and competing synonyms, it cannot be conserved against "itself".

At present there are two appendices with conserved family names in the *Code*. One of them is in a way rather irregular and, until recently, was not covered by the *Code*'s own provisions on conservation: Appendix IIB, in which no rejected names are listed against the conserved ones. The present Appendix IIB is in fact little else than a list of protected names as envisaged in the new proposals. There are a few cases of genuine conservation in it, consigned to "Notes" stating that when two families bearing conserved names are united the younger name is to be used. The new proposals would effect that the Appendix IIB situation would be covered by protection of the corresponding approved lists, so that Appendix IIB could be deleted and the few cases of real conservation transferred to Appendix IIA. This would concern only family names of Spermatophyta and Bryophyta. Since the new lists include many names not now in Appendix IIB, some of which would, under reasonable taxonomic options, threaten names now listed as conserved, some further conservations would be needed that have been foreshadowed here by the words "to be conserved against ..." or "to be rejected in favour of ...". The conserved names of Pteridophyta and Fungi now in Appendix IIA would remain, but many of the corresponding rejected names (those not listed as being in current use) could be deleted.

Spelling of family names

When one looks up the protologues one finds that original spellings of family names vary widely and often differ considerably from the spellings given in the following lists. Nevertheless there is only one correct spelling for each family name, and all other spellings, according to the *Code,* are treated as not being validly published. There is a special rule in the *Code* mandating the correction of a wrong termination originally used for a family name. The general rule on correctability of orthographic errors applies when a family name has been based on an incorrect spelling of the generic name from which it is derived, or on an incorrectly formed genitive singular of it.

Correction of spelling does not result in a change of the original authorship and place of publication of a name. This is fortunate because, for older family names in particular, it is rather exceptional to find that the original spelling is the one now considered as correct. Making mention of the original, incorrect spellings, now devalidated by law, is a spurious exercise of purely academic interest, and we have therefore eliminated such citations although they still appear, for historical reasons, in the lists of conserved family names in Appendix II of the *Code.*

The spelling of conserved (and protected) names is (or could be) treated as conserved. This has some importance in cases in which there may be doubt as to the correct spelling to be adopted. Uniformization of spelling is gaining increased importance in a time when electronic storage and retrieval of data is omnipresent, since a consistent spelling is a prerequisite of easy access to the data. For a family name the correct spelling depends on two variables: the spelling of the generic name on which it is based, and that of its genitive singular. It is the latter that has mainly created problems, which are peculiar to names at suprageneric ranks. We have devoted particular attention to this aspect when compiling the lists to follow. Since the same questions will arise with names of subdivisions of families, not treated here, it may be of some relevance to explain in detail how we have proceeded.

For most Latin nouns and a majority of Greek ones, the genitive singular is isosyllabic ("parisyllabic"), i.e., it has the same number of syllables as the nominative singular. Such names do not present problems: you just chop off the termination (normally: *-us* or *-os, -a* or *-e, -um* or *-on, -is* or *-es*) and add the family termination *-aceae* instead. Exceptions with one of the above endings but anisosyllabic ("imparisyllabic") genitive singular do occur, mostly in Greek. When listing them here I will not attempt encyclopaedic coverage but will limit myself to examples found among the listed names.

Nouns of neuter gender ending in *-ma* have a genitive singular stem *-mat-* and yield family names ending in *-mataceae.* The generic name *Collema* and those compounded with *derma, gramma* (the letter), *loma, nema, omma, phorema, phragma, soma, sperma, stema, stigma, stoma, stroma, trema* or *zoma,* or ending in *-asma, -isma* or *-oma,* belong to this category. In contrast, feminine nouns ending in *-ma,* either truly Latin (as *forma, coma*) or Latinized from Greek *-me (gramme* the line, *kneme, nome, osme, parme)* have isosyllabic genitives and yield family names ending in *-maceae.* The same applies to the feminine generic names *Dirachma, Phryma,* and *Rhynchogastrema,* the etymological source of which is unclear but which were treated as feminine by their authors. One may note the special case of names compounded with *gramma,* some of which are feminine and some neuter: this is our only instance in which one and the same terminal element of a generic name may give different family name terminations.

The only exception to the *-us* rule, among our examples, is a compound of the Greek *pus* (foot), with corresponding family name ending *-podaceae.*

Whereas Greek neuter nouns ending in *-on* invariably have regular, isosyllabic genitive formation, a few masculine nouns with that ending *(geton, pogon, siphon, stemon)* have anisosyllabic declension and yield family names ending in *-onaceae.*

The *-is* ending of Greek or Latin words in botanical tradition almost invariably results in family names ending in *-idaceae* (anisosyllabic), even in those cases in which classical usage is different, as among others in the Greek *orchis* and *opsis,* or when it varies. The only isosyllabic exceptions we found are compounds with *axis, crinis,* and *pellis,* as well as the generic names *Acrasis, Batis, Cannabis, Capparis, Fontinalis, Haloragis, Helosis,* and *Vitis.* Three or four of these are, by the way, conserved spellings which by following botanical tradition would have been treated differently. A second set of exceptions are compounds with the Greek *charis* and *margaris* which result in family names ending in *-itaceae.*

The *-es* ending, on the contrary, almost universally belongs to regularly declined names with an isosyllabic genitive. *Abies* (genitive *Abietis*) having been removed from the lists, the only remaining exception are compounds of *myces* with corresponding family name termination *-mycetaceae.*

The other, non-standard endings of generic names are much rarer but all have their particularities. The following have been encountered and will be briefly discussed here: *-er, -as, -ys, -o, -x,* and consonant plus *-s.*

All relevant cases of *-er* ending have anisosyllabic declension and give *-eraceae* family name terminations, such as the generic names *Papaver, Piper,* and *Zingiber,* although two pose problems: *gaster* has a dual genitive singular *(gastri/gasteris)* of which the second only has been used here for the sake of consistency; *aster* meaning star has a genitive *asteris* but the pejorative suffix *-aster* is declined *-astri* and would yield *-astraceae.* Both etymologies are found but in our single instance of the second type *(Circaeaster)* the family name is already conserved with the grammatically incorrect spelling *Circaeasteraceae.*

Names in *-as (Asclepias, Cycas, Najas,* and compounds of *thylas)* form family names in *-adaceae,* except for compounds of *ceras* which result in *-cerataceae.* The *-ys* ending usually *(botrys, pitys, stachys)* yields *-yaceae* but exceptionally *(chlamys) -ydaceae.*

The *-o* ending has been met with in the two Latin suffixes *-ago* and *-ugo,* both of which give family names in *-ginaceae;* in compounds of *umbo* and in the generic name *Restio,* leading to *-onaceae.*

Names ending in *-x* (the phonetic equivalent of *-ks*) normally result in *-caceae* family name endings, with two notable exceptions: *mastix* takes a *g* in the genitive (hence *-mastigaceae*) and the highly irregular *thrix* yields *-trichaceae.* To complete this inventory, let us mention names compounded with *ops (-opaceae), ceps (-cipitaceae),* and *glans (-glandaceae).*

Last, we found three generic names which are neither Greek nor Latin by any standard, so that the family name ending *-aceae* is added to the full word: *Caryocar, Ginkgo* and, for cross-reference purposes, *Ribes.*

Author citations

Citing the authors of a scientific plant name is not quite as easy and straightforward a matter as it might appear. Neither the question of which authors to nor that of how to cite them exactly is an easy one. No less than 34 proposals to modify the rules and recommendations on author citation in the *Code* are presently under consideration, more than 10 % of the proposals submitted to the Tokyo Congress. Although the sheer number of proposals may be taken as a forceful omen for no change in substance to happen in Tokyo, it also doubtless indicates the existence of problems and uncertainty.

For the purpose of the following lists we have made a neat distinction between authors in the nomenclatural sense and authors of publications. We have followed the recommendations of the Special Committee on "in" and "ex" Citations in considering that the particle "in" and the personal names that may follow are not part of the nomenclatural author citation but of the bibliographic reference, and have not therefore abbreviated such personal names but cite them in full, in conformity with standard bibliographic procedure. Personal names that precede the "in" are, on the other hand, authors in the nomenclatural sense. The spelling of their names, their abbreviation if applicable, and the use of initials when appropriate, follow the new internationally recommended standard: Brummitt & Powell's book *Authors of plant names.* In those cases when the author(s) of the name and the author(s) of the validating publication are exactly the same they are, obviously, cited only once, and the nomenclatural author standard is then followed; this is the situation when the particle "in" disappears and is replaced by a comma.

Particles may occur also within the author citation proper. Joint authorship is indicated by the ampersand (&) that stands for Latin "et" (and); of three or more co-authors only the first is named, followed by "& al." Alleged authorship (attribution of a name, by its real author(s), to another authority) is mentioned ahead of the true authorship, both being separated by the particle "ex", it being understood that citation of this particle and the preceding name(s) it is optional. It is not always easy to distinguish between true and alleged authorship, but we have tried to follow the rules and botanical tradition as best we could. Pre-starting-point authors have, in principle, not been cited.

Parenthetical authorship citations after family names are a particularly controversial topic. They are sometimes used, even in the *Code,* but there is no clear provision on which such use could be based. Proposals to eliminate parethetical authorship citations after suprageneric names having failed at recent Congresses, their use is obviously authorized, but we consider it as optional. The criteria applied for adding parenthetical authorship are explained in greater detail by Hoogland & Reveal in their preface to the Vascular Plant lists. A further relevant point is mentioned below, under the heading "Basionyms".

Dates and places of valid publication

Great care has been devoted to the consistent citation of the bibliographical sources of all names. Needless to say, all these sources were verified by the authors of the lists, so that the references given can be taken as accurate and fulfilment of the conditions of valid publication can be vouched for. Bibliographic citations have been abbreviated in conformity with internationally recognized standards, in the following way.

For journals and serials, the abbreviations listed in the *Botanico-Periodicum Huntianum* (Pittsburgh, 1968) and its Supplement (Pittsburgh, 1991) have been used, allowing for a few matter-of-fact corrections. While changes introduced in the Supplement have been accepted in general, an exception has been made for the transliteration of Cyrillic characters for which we continue to prefer the ISO-1 system to ISO-2, be it only in fairness to those Slavic languages which, like Czech, Slovenian and Croatian, use the Latin alphabet with ISO-1-type diacritic marks. All journal and anonymous serial titles are preceded by the particle "in".

The titles of books and other independent publications have been abbreviated in conformity with Stafleu & Cowan's *Taxonomic literature,* ed. 2, and its Supplement, or by analogy when not included there. Titles in non-Latin script have been transliterated using internationally agreed standards, and in some cases an abbreviated English translation has been added in brackets for convenience.

For many names it is customary to give a dual citation of their place of valid publication. This may have various reasons. The two citations may actually refer to the same publication that is known under different titles; or what is basically the same publication may have been published twice, once in a journal and once independently (or in two different journals, or in two different editions), with either the same or a different pagination, either simultaneously or at different dates. This poses a problem when it is envisaged to grant protected status to the cited place and date of publication of listed names.

We have as a matter of principle given preference to a single citation, although for clarity and convenience we have often added a second, bracketed reference. In so doing we have always preferred the journal citation to the citation from an independent publication, except when the latter is both earlier and independently paged. Preprints bearing the journal pagination have been considered as being part of the journal published on the earlier date. Of two different, simultaneous editions the more widespread and more easily accessible one was given preference.

Within each bibliographic citation, a colon (:) separates the elements that refer to the whole publication (title, edition, series, volume and fascicle number, as appropriate) from those relevant for the individual name (page and/or plate); and a period (.) separates the latter from the date of publication.

The dates of publication have been given as exactly as possible. Full use has been made of the information included in *Taxonomic literature,* ed. 2, and in other scattered sources. Printed dates appearing in the publication itself have been accepted as correct unless they are demonstrably erroneous. The search for datings may not have been quite as exhaustive as we might have wished, but should be amply sufficient for establishing the relative priority of potentially competing listed names.

Basionyms

"Basionyms" have been given, with full details of their source of validation, whenever parenthetical authors of family names have been cited. This requires some comment since, although it conforms to traditional usage, it is not covered by the provisions of the *Code*. Currently a basionym is defined as the "name-bringing or epithet-bringing synonym" of a name. Epithets are out of the question at suprageneric ranks, and what we call basionyms here are not "name-bringing" synonyms either. If anything, the name-bringing synonym hence the basionym of a family name, according to this definition, would be the generic name on which it is based. But then, reference to a validly published, name-bringing generic name is not sufficient to validate the name of a new family.

The question may legitimately be asked whether full and direct reference to a validly published name of a subdivision of a family is nowadays sufficient to validate a new family name, even though no Latin description or diagnosis may be involved. In other words: are we always dealing with a new taxon when naming a hitherto unnamed family? Or may we just transfer a pre-existing, validly named subdivision of a family to family rank and bypass the Latin requirement? And, granting the latter possibility, may we in so doing change the circumscription of the taxon? The *Code* is remarkably silent on these questions, nowhere defining what a "new taxon" is, but we are satisfied that the approach we have chosen (see also Hoogland & Reveal's Introduction) is the one that best reflects what has traditionally been understood.

Nomenclatural types

The name of a family is typified by the type of the name of the genus on which it is based. No attempt has been made in the present lists to establish types beyond listing the generic names in question. Verification and eventually stabilization of the types of generic names must occur at another level and need not concern us here.

Most generic names cited under the heading "Type" are familiar, but some are no longer in current use and are not included in the generic NCU list that is presently being prepared. Some of them are homotypic with older or nomenclaturally conserved generic names, either through subsequent lectotypification or by being illegitimate renamings; in these cases their nomenclatural equivalents are added in parentheses. Others have long gone out of use as taxonomic synonyms of earlier names, in which case the earlier name is added in brackets, preceded by an equal sign. (By a convention familiar to the users of Appendix III of the *Code,* parentheses always include homotypic synonyms, and square brackets, heterotypic ones; see Taxon 33: 311. 1984.) The synonymies are given only for information and are in no way designed to have nomenclatural implications.

Should the types of the listed names be granted protected status, as would be possible under the proposed principles of nomenclatural protection, this would be of little consequence generally speaking. It would, however, make it explicit that the spelling not only of the family names but also of the generic names that stand for their types is protected as listed.

Names with equal priority

Four pairs and two triplets of currently used vascular plant family names that were published simultaneously in the same work may, under a reasonable taxonomic assumption, be considered to be synonyms. In such situations the NCU proposals would provide for the possibility of a firm ruling as to which name is to be treated as having priority over the other(s). This has been foreshadowed, to become effective if and when the lists are nomenclaturally approved (see Hoogland & Reveal's Introduction), so that *Lophosoriaceae* would then be sure to prevail over *Metaxyaceae, Cornaceae* over *Nyssaceae, Loganiaceae* over *Potaliaceae* and *Spigeliaceae, Sapindaceae* over *Aceraceae, Liliaceae* over *Asparagaceae* and *Asphodelaceae,* and *Xanthorrhoeaceae* over *Dasypogonaceae.*

Acknowledgements

It would not have been possible to produce this inventory in a relatively short time and in such a well digested and thoroughly checked form without the devoted cooperation of the authors of the individual lists. Thanks are due to them all, and most particularly to Ru Hoogland, Gea Zijlstra and John David who were my immediate partners during the process of editing. They had to bear with many changes they did not lightly approve but which were necessary for achieving consistency of style and presentation between the different lists, and between them and the generic NCU list.

During the early stages of preparing the lists, standardization progress was hampered by the fact that both the new author abbreviation standard by Brummitt & Powell and the Supplement to the *Botanico-Periodicum Huntianum* were not yet available in printed form. Thanks to a small grant by IUBS and IAPT, and under the supervision of Dick Brummitt, Lulu Rico at Kew has undertaken to verify a long list of author names against the then still unpublished text files of *Authors of plant names.* The Director of the Royal Botanic Gardens, Kew, and Bob Kiger of the Hunt Institute for Botanical Documentation deserve our sincere thanks for having put at our disposal the corresponding texts in electronic form, which has greatly facilitated our more recent checks. I owe a great debt of gratitude to Brigitte Zimmer, Berlin, for having produced camera-ready copy of this whole boocklet.

The Permanent Committees for Spermatophyta, Bryophyta, and Fungi and Lichens have considered the relevant draft lists and are to be thanked for their helpful attitude and constructive feedback. Critical evaluation of both the early draft and the new printed version of the lists has been and will be greatly appreciated.

Comments have been received by numerous correspondents who will, we hope, find that their suggestions have been duly considered and in most cases accepted. Among them Rodolfo E. G. Pichi-Sermolli (Italy) stands out by his very detailed and competent contribution to pteridophyte family nomenclature. All others also deserve our sincere thanks for their interest and help: André Aptroot (Netherlands), Günther Buchheim (U.S.A.), Aljos Farjon (Netherlands), David L. Hawksworth (U.K.), Surrey Jacobs (Australia), Makoto Kakishima (Japan), Takao Kobayashi (Japan), Richard P. Korf (U.S.A.), Hermann Neubert (Germany), Gen. Okada (Japan), Y. Ono (Japan), Erast Parmasto (Estonia), David N. Pegler (U.K.), Bernard Pelletier (Canada), Zdeněk Pouzar (Czechia), Guy Redeuilh (France), Junta Sugiyama (Japan), John Walker (Australia), Karen L. Wilson (Australia), and Kazumasa Yokoyama (Japan).

VASCULAR PLANT FAMILY NAMES IN CURRENT USE

By Ruurd D. Hoogland and James L. Reveal

The list presented here reflects current use of family names of vascular plants. If the XV International Botanical Congress in Tokyo accepts the proposed principle of granting nomenclatural protection to listed names in current use (Greuter, 1991), and if the present list is indeed approved, the names listed here will be treated as if conserved against earlier homonyms and unlisted synonyms, and as validly published in the place and on the date cited in the list.

We have included those family names that have been used for families taxonomically accepted in one or more of a number of major texts and reference works published since 1970 (Airy Shaw in Willis, 1973; Bedell & Reveal, 1982a, b; Brummitt, 1992; Ching, 1978; Crabbe & al., 1975; Cronquist, 1981, 1988; G. Dahlgren, 1989a, b; R. Dahlgren, 1975, 1980, 1983; R. Dahlgren & Bremer, 1985; R. Dahlgren & Clifford, 1982; R. Dahlgren & al., 1981, 1985; Goldberg, 1986, 1989; Green in Kubitzki, 1990; Gunn & al., 1992; Heywood, 1974; Hutchinson, 1973; Kramer in Kubitzki, 1990; Mabberley, 1987; Pichi Sermolli, 1977, 1981; Rouleau, 1970, 1981; Stebbins, 1974; Takhtajan, 1970, 1973, 1980, 1983, 1985, 1986, 1987; Thorne, 1974, 1976, 1981, 1983, 1992a, b; A. Tryon & Lugardon, 1990; R. Tryon & A. Tryon, 1982); in determining use of names we have worked closely with Cronquist, Takhtajan and Thorne to resolve those instances where priority would overturn established names. All currently conserved family names, even if not accepted by any of the above authors, have been included. Some families, not accepted in any of the listed publications but accepted in one or more other publications since 1970, are also included. Finally a number of family names published since 1970, hence not taken into consideration by all or even any of the listed authors, have been added.

Although the *Code* (Greuter & al., 1988) is ambiguous in this respect, we have accepted the principle that a family name can be validated by reference to an earlier, validly published name of a subdivision of a family, even though no Latin description or diagnosis may be involved. The validity of such transfers may admittedly be challenged, which would however run against the interests of nomenclatural stability and be at variance with traditional nomenclatural practice. Such uncertainty over the validity of names that are needed is a serious drawback for nomenclatural stability, and a major argument in support of protection of the listed names.

In our list we have given, where appropriate, the reference to a "basionym" and have cited the authority for this basionym in parentheses before the authority for the family name, irrespective of whether the protologue has its own validating description or diagnosis in addition to the reference to a previously and validly published name of a subdivision of a family (e.g., Airy Shaw, 1965), or merely such a reference (e.g., Takhtajan, 1987), and whether the validating author used parenthetical authorities after the family name, as Airy Shaw did, or not, as in Takhtajan (but not when a potential validly published "basionym" merely exists, without being cited). The *Code* is again inconclusive as to the appropriateness of citing parenthetically the author(s) of a validly published "basionym" at a rank of subdivision of a family, and the way in which such situations are presented in its examples and appendices is inconsistent (App. IIA includes some parenthetical author citations for family names, whereas they have been consistently omitted in App. IIB in comparable situations).

For a substantial (over a hundred) number of presently conserved family names we found earlier places of valid publication which under the Berlin *Code* would have to replace the currently listed reference. In addition we have found a number of cases where the given reference is to a publication in which, in our judgement, the conditions for valid publication of the name are not fulfilled, and would have to be replaced by reference to a later publication where these are fulfilled. The NCU proposals would make it possible that, once this list is approved, the references currently given in the *Code* (and in the list below) for these names be retained, an option that does not exist under the present provisions for conservation. If the NCU proposals are not or only partly accepted, it will be necessary to replace the references for at least a quarter of the names in App. IIB, and even so one will not eliminate the risk of future changes should yet earlier valid publications for some family names be found, as is likely.

For one of the family names included in App. IIA of the *Code (Adiantaceae)* we have accepted an earlier reference than the one currently given, as recommended by the Committee for Pteridophyta (Lellinger, 1993). For family names presently listed in App. IIB we have in principle not changed the reference except for trivial editorial corrections and the addition of parenthetical author citations and "basionyms" when appropriate.

Proposed changes (Reveal & Hoogland, 1992b) to entries of presently conserved family names in App. IIB of the Berlin *Code,* except the purely editorial ones, have not been effected in view of lack of support for them by the Committee for Spermatophyta (Brummitt, pers. comm.). For similar reasons, only four of the family names for which we proposed conservation (Reveal & Hoogland, 1992a) have been included. For the other four, in two cases a legitimate name either existed or has been validated, and those are now listed (*Oftiaceae* Takht. & Reveal, *Phyllonomaceae* Small). In a further case those who wish to recognize the family will have to publish a new name since the existing one is illegitimate (*Aitoniaceae* Harv.). Finally the name *Abietaceae* has been removed from the list as not being in current use, which leaves the unlisted, older name *Cedraceae* Vest available for use by those who would wish to restrict the family *Pinaceae* to the genus *Pinus.*

For family names not currently conserved we have given the reference to the earliest place of valid publication that we have been able to find, using Jussieu (1789) as the starting point book and date. In this we follow Bullock (1958) and Buchheim (1966) as well as the list of conserved family names, but ignore Buchheim's statement that the normal 1753 starting point will hold for all names that are not conserved. Undoubtedly still earlier (post-1789) references can be dug up for some of these names. However, if the NCU proposals are accepted, the authorship and place of publication given here can be protected.

A number of names which were published by Endlicher (1836-1841, 1843) without clear indication of rank have traditionally been accepted as family names, and twelve of them are currently included in App. IIB. These are names which Endlicher mentioned at the end of some of his accepted families ("ordines"). We have followed tradition and have included a further nine similarly published family names from these two sources.

In the present list names that are currently included in App. IIB have all been given as "*nom. cons.*" With a few exceptions, their status will change to that of protected names upon acceptance of the NCU proposals and approval of the list. This means that they will lose their conserved status against names not previously conserved but now listed. In order to avoid undesirable name changes for broadly circumscribed families, it will be necessary to grant explicit conservation to a few additional family names. Such additional entries for the future App. II have been foreshadowed by adding statements such as "to be conserved against ..." and "to be rejected in favour of ..." after the relevant entries of our list. Furthermore, in cases of equal priority of potentially competing names a statement has been added to make clear which name, when the situation arises, "is treated as having priority". Note that the rule: "For any family including the type of an alternative family name, one or the other of these alternative names is to be used", now in App. IIB of the *Code,* will continue to be relevant.

Our work on family names continues. We hope to produce a catalogue of all vascular plant family names in 1995, and will appreciate any information or comments on family names readers would care to send.

Acknowledgements

Several colleagues, libraries and botanical institutions have kindly assisted us in improving and editing the version that was distributed in August 1991. It is impossible to list them all here, but we wish to thank in particular Günther Buchheim, the late Arthur J. Cronquist, Aljos K. Farjon, Werner Greuter, Rodolfo E. G. Pichi Sermolli, Armen L. Takhtajan, Robert

F. Thorne and Gea Zijlstra. The first author wishes to express his gratitude to the Laboratoire de Phanérogamie (Paris) for providing working facilities and other assistance. The second author's work on plant family nomenclature was supported, in part, by National Science Foundation Grant BSR-8812816. [This is Scientific Article A-6396, Contribution No. 8583, of the Maryland Agricultural Experiment Station and Cooperative Extension Service.]

Literature cited

Airy Shaw, H. K. 1965. Diagnoses of new families, new names, etc. for the seventh edition of Willis's 'Dictionary'. *Kew Bull.* 18: 249-273.

Bedell, H. G. & Reveal, J. L. 1982a. Amended outlines and indices for six recently published systems of angiosperm classification. *Phytologia* 51: 65-156.

– & – 1982b. A synoptical review of a revised classification of Liliopsida (Magnoliophyta) as proposed by Dahlgren and Clifford. *Phytologia* 52: 179-183.

Brummitt, R. K. 1992. *Vascular plant families and genera.* Kew.

Buchheim, G. 1966. Report of the Subcommittee for family names of the Committee for Spermatophyta. [*In:* Stafleu, F. A., Tenth International Botanical Congress 1964. Nomenclature Section.] *Regnum Veg.* 44: 61-65.

Bullock, A. A. 1958. Indicis nominum familiarum Angiospermarum prodromus. *Taxon* 7: 1-35.

Ching, R.-C. 1978. The Chinese fern families and genera: systematic arrangement and historical origin. *Acta Phytotax. Sin.* 16(3): 1-19; 16(4): 16-37.

Crabbe, J. A., Jermy, A. C. & Mickel, J. T. 1975. A new generic sequence for the pteridophyte herbarium. *Fern Gaz.* 11: 141-162.

Cronquist, A. J. 1981. *An integrated system of classification of flowering plants.* New York.

– 1988. *The evolution and classification of flowering plants,* ed. 2. Bronx, NY.

Dahlgren, G. 1989a. An updated angiosperm classification. *Bot. J. Linn. Soc.* 100: 197-203.

– 1989b. The last Dahlgrenogram. System of classification of the dicotyledons. Pp. 237-248 *in:* Tan, K., Mill, R. R. & Elias, T. S. (ed.), *Plant taxonomy, phytogeography and related subjects: The Davis and Hedge festschrift.* Edinburgh.

Dahlgren, R. M. T. 1975. A system of classification of the angiosperms to be used to demonstrate the distribution of characters. *Bot. Not.* 128: 119-147.

– 1980. A revised system of classification of the angiosperms. *Bot. J. Linn. Soc.* 80: 91-124.

– 1983. General aspects of angiosperm evolution and macrosystematics. *Nordic J. Bot.* 3: 119-149.

– & Bremer, K. 1985. Major clades of the angiosperms. *Cladistics* 1: 349-368.

– & Clifford, H. T. 1982. The monocotyledons: a comparative study. *Bot. Syst.* 2.

– , – & Yeo, P. F. 1985. *The families of the monocotyledons. Structure, evolution, and taxonomy.* Berlin, etc.

– , Rodendal-Jensen, S. & Nielsen, B. J. 1981. A revised classification of the angiosperms with comments on correlation between chemical and other characters. Pp. 149-204 *in:* Young, D. A. & Seigler, D. S. (ed.), *Phytochemistry and angiosperm phylogeny.* New York.

Endlicher, S. F. L. 1836-1841. *Genera plantarum secundum ordines naturales disposita.* Wien.

– 1843. *Enchiridion botanicum exhibens classes et ordines plantarum.* Leipzig & Wien.

Goldberg, A. 1986. Classification, evolution, and phylogeny of the families of dicotyledons. *Smithsonian Contr. Bot.* 58.

– 1989. Classification, evolution, and phylogeny of the families of monocotyledons. *Smithsonian Contr. Bot.* 71.

Greuter, W. 1991. Proposals to amend the Code and report of Special Committee 6B: Provisions for granting nomenclatural protection to listed names in current use. *Taxon* 40: 669-677.

– , Burdet, H. M., Chaloner, W. G., Demoulin, V., Grolle, R., Hawksworth, D. L., Nicolson, D. H., Silva, P. C., Stafleu, F. A., Voss, E. G. & McNeill, J. (ed.), 1988. International code of botanical nomenclature, adopted by the Fourteenth International Botanical Congress, Berlin, July-August 1987. *Regnum Veg.* 118.

Gunn, C. R., Wiersema, J. H., Ritchie, C. A. & Kirkbride, Jr., J. H. 1992. Families and genera of spermatophytes recognized by the Agricultural Research Service. *U.S.D.A. Techn. Bull.* 1796.

Heywood, V. H. 1978. *Flowering plants of the world.* Oxford & New York. [Reprinted without change in 1985.]

Hutchinson, J. 1973. *The families of flowering plants arranged according to a new system based on their probable phylogeny* , ed. 3. Oxford.

Kubitzki, K. 1990. *The families and genera of vascular plants. I. Pteridophytes and gymnosperms.* Berlin, etc.

Lellinger, D. B. 1993. Report of the Committee for Pteridophyta: 13. *Taxon* 42: 119-120.

Mabberley, D. J. 1987. *The plant-book. A portable dictionary of the higher plants.* Cambridge, etc. [Reprinted in 1989.]

Pichi Sermolli, R. E. G. 1977. Tentamen pteridophytorum genera in taxonomicum ordinem redigendi. *Webbia* 31: 313-512.

– 1981. Report of the subcommittee for family names of Pteridophyta. *Taxon* 30: 163-168.

Reveal, J. L. & Hoogland, R. D. 1992a. Proposals to conserve eight family names of Spermatophyta. *Taxon* 41: 116-121.

– & – 1992b. Proposals to modify entries in Appendix IIB of the Code. *Taxon* 41: 122-125.

Rouleau, E. 1970. *Guide to index kewensis and its supplements i-xiv.* Montreal.

– 1981. *Répertoire des noms génériques mentionnés dans l'index kewensis et ses quinze premiers suppléments.* Lac de Brome.

Stebbins, G. L. 1974. *Flowering plants. Evolution above the species level.* Cambridge MA & London.

Takhtajan, A. L. 1970. *Proishoždenie i rasselenie cvetkovyh rastenii.* [Origin and dispersal of the flowering plants.] Leningrad.

– 1973. *Evolution und Ausbreitung der Blütenpflanzen.* Stuttgart.

– 1980. Outline of the classification of flowering plants (Magnoliophyta). *Bot. Rev. (Lancaster)* 46: 225-359.

– 1983. The systematic arrangement of dicotyledonous families. Pp. 180-201, 267-273 *in:* Metcalfe, C. R. & Chalk, L., *Anatomy of the dicotyledons,* ed. 2, 2. Oxford.

– 1985. *Sravnitel'naja anatomija semjan* [Anatomia seminum comparativa], 1. Leningrad.

– 1986. *Floristic regions of the world.* Berkeley, etc.

– 1987. *Sistema magnoliofitov.* [Systema Magnoliophytorum.] Leningrad.

Thorne, R. F. 1974. A phylogenetic classification of the Annoniflorae. *Aliso* 8: 147-209.

– 1976. A phylogenetic classification of the Angiospermae. *Evol. Biol.* 9: 35-106.

– 1981. Phytochemistry and angiosperm phylogeny: a summary statement. Pp. 233-295 *in:* Young, D. A. & Seigler, D. S. (ed.), *Phytochemistry and angiosperm phylogeny.* New York.

– 1983. Proposed new realignments in the angiosperms. *Nordic J. Bot.* 3: 85-117.

– 1992a. An updated phylogenetic classification of the flowering plants. *Aliso* 13: 365-389.

– 1992b. Classification and geography of the flowering plants. *Bot. Rev. (Lancaster)* 58: 225-348.

Tryon, A. F. & Lugardon, B. 1990. *Spores of the Pteridophyta. Surface, wall structure, and diversity based on electron microscope studies.* New York, etc.

Tryon, R. M. & Tryon, A. F. 1982. *Ferns and allied plants. With special reference to tropical America.* New York, etc.

Willis, J. C. 1973. *A dictionary of the flowering plants and ferns,* ed. 8. Cambridge, etc. [Reprinted without change in 1980, 1985, 1988.]

Pteridophyta

Acrostichaceae Mett. ex A. B. Frank in Leunis, Syn. Pflanzenk., ed. 2: 1453, 1458. Mar-Dec 1877. Type: *Acrostichum* L.

Actiniopteridaceae Pic. Serm. in Webbia 17: 5. 31 Aug 1962. Type: *Actiniopteris* Link

Adiantaceae Newman, Hist. Brit. Ferns: 5. 1-5 Feb 1840, *nom. cons.* (conserved against *Parkeriaceae*). Type: *Adiantum* L.

Anemiaceae Link, Handbuch 3: 8. ante 21 Apr 1833. Type: *Anemia* Sw., *nom. cons.*

Angiopteridaceae Fée ex J. Bommer [Monogr. Foug.: 87, 92] in Bull. Soc. Roy. Bot. Belgique 5: 345, 359. ante 28 Mai 1867. Type: *Angiopteris* Hoffm., *nom. cons.*

Antrophyaceae (Link) Ching in Acta Phytotax. Sin. 16(3): 11. Aug 1978. Basionym: subfam. *Antrophyoideae* Link, Fil. Spec.: 140. 3-10 Sep 1841. Type: *Antrophyum* Kaulf.

Aspidiaceae, nom. illeg. – see *Tectariaceae*

Aspleniaceae Newman, Hist. Brit. Ferns: 6. 1-5 Feb 1840. Type: *Asplenium* L.

Athyriaceae Alston in Taxon 5: 25. 25 Mar 1956. Type: *Athyrium* Roth

Azollaceae Wettst., Handb. Syst. Bot. 2: 77. Nov 1903. Type: *Azolla* Lam.

Blechnaceae (C. Presl) Copel. [Gen. Fil.] in Ann. Cryptog. Phytopathol. 5: 155. medio Sep 1947. Basionym: trib. *Blechneae* C. Presl [Epimel. Bot.: 103] in Abh. Königl. Böhm. Ges. Wiss., ser. 5, 6: 463. Oct 1851. Type: *Blechnum* L.

Bolbitidaceae (Pic. Serm.) Ching in Acta Phytotax. Sin. 16(3): 15. Aug 1978. Basionym: trib. *Bolbitideae* Pic. Serm. in Webbia 23: 381. 30 Jun 1969. Type: *Bolbitis* Schott

Botrychiaceae Horan., Char. Ess. Fam.: 15. 1847. Type: *Botrychium* Sw.

Cheilanthaceae B. K. Nayar in Taxon 19: 233. 29 Apr 1970. Type: *Cheilanthes* Sw., *nom. cons.*

Cheiropleuriaceae Nakai in Bot. Mag. (Tokyo) 42: 210. Apr 1928. Type: *Cheiropleuria* C. Presl

Christenseniaceae Ching in Bull. Fan Mem. Inst. Biol., Bot. 10: 227. 25 Dec 1940. Type: *Christensenia* Maxon

Cryptogrammaceae Pic. Serm. in Webbia 17: 299. 20 Apr 1963. Type: *Cryptogramma* R. Br.

Culcitaceae Pic. Serm. in Webbia 24: 702. 28 Apr 1970. Type: *Culcita* C. Presl

Cyatheaceae Kaulf., Wesen Farrenkr.: [119]. Jul-Sep 1827. Type: *Cyathea* Sm.

Cystodiaceae J. R. Croft in Kew Bull. 41: 797. 20 Oct 1986. Type: *Cystodium* J. Sm., *nom. cons.*

Danaeaceae C. Agardh, Aphor. Bot.: 117. 19 Jun 1822. Type: *Danaea* Sm., *nom. cons.*

Davalliaceae Mett. ex A. B. Frank in Leunis, Syn. Pflanzenk., ed. 2: 1453, 1474. Mar-Dec 1877. Type: *Davallia* Sm.

Dennstaedtiaceae Lotsy, Vortr. Bot. Stammesgesch. 2: 655. Jul-Sep 1909. Type: *Dennstaedtia* Bernh.

Dicksoniaceae (C. Presl) Bower, Origin Land Fl.: 591. 1908, *nom. cons.* (conserved against *Thyrsopteridaceae*). Basionym: trib. *Dicksonieae* C. Presl [Tent. Pterid.] in Abh. Königl. Böhm. Ges. Wiss., ser. 4, 5: 133. [seors. impr.] ante 2 Dec 1836. Type: *Dicksonia* L'Hér.

Dipteridaceae (Diels) Seward & E. Dale in Philos. Trans., Ser. B, 194: 487, 499, 502. Dec 1901. Basionym: subtrib. *Dipteridinae* Diels in Engler & Prantl, Nat. Pflanzenfam. 1(4): 167, 202. Aug 1899. Type: *Dipteris* Reinw.

Drynariaceae Ching in Acta Phytotax. Sin. 16(4): 19. Nov 1978. Type: *Drynaria* (Bory) J. Sm., *nom. cons.*

Dryopteridaceae Ching in Acta Phytotax. Sin. 10: 1. Jan 1965, *nom. cons.* (conserved against *Peranemataceae;* to be conserved against *Woodsiaceae*). Type: *Dryopteris* Adans., *nom. cons.*

Elaphoglossaceae Pic. Serm. in Webbia 23: 209. 20 Nov 1968. Type: *Elaphoglossum* Schott ex J. Sm., *nom. cons.*

Equisetaceae Michx. ex DC., Essai Propr. Méd. Pl.: 49. Jul 1804. Type: *Equisetum* L.

Gleicheniaceae (R. Br.) C. Presl, Reliq. Haenk. 1: 70. Jun-Nov 1825. Basionym: [unranked] *Gleicheneae* R. Br., Prodr.: 160. 27 Mar 1810. Type: *Gleichenia* Sm., *nom. cons.*

Grammitidaceae Newman, Hist. Brit. Ferns: 7. 1-5 Feb 1840. Type: *Grammitis* Sw.

Gymnogrammaceae, nom. illeg. – see *Hemionitidaceae*

Gymnogrammitidaceae Ching in Acta Phytotax. Sin. 11: 12. Jan 1966. Type: *Gymnogrammitis* Griff.

Helminthostachyaceae Ching in Bull. Fan Mem. Inst. Biol., Bot.

10: 235. 15 Mar 1941. Type: *Helminthostachys* Kaulf.

Hemionitidaceae Pic. Serm. in Webbia 21: 487. 15 Jul 1966. Type: *Hemionitis* L.

Huperziaceae Rothm. in Feddes Repert. Spec. Nov. Regni Veg. 66: 236. 15 Nov 1962. Type: *Huperzia* Bernh.

Hymenophyllaceae Link, Handbuch 3: 36. ante 21 Apr 1833. Type: *Hymenophyllum* Sm.

Hymenophyllopsidaceae Pic. Serm. in Webbia 24: 712. 28 Apr 1970. Type: *Hymenophyllopsis* K. I. Goebel

Hypodematiaceae Ching in Acta Phytotax. Sin. 13(1): 96. Jan 1975. Type: *Hypodematium* Kunze

Hypolepidaceae Pic. Serm. in Webbia 24: 705. 28 Apr 1970. Type: *Hypolepis* Bernh.

Isoëtaceae Rchb., Bot. Damen: 309. 1828. Type: *Isoëtes* L.

Kaulfussiaceae, nom. illeg. – see *Christenseniaceae*

Lindsaeaceae Pic. Serm. in Webbia 24: 707. 28 Apr 1970. Type: *Lindsaea* Dryand. ex Sm.

Lomariopsidaceae Alston in Taxon 5: 25. 25 Mar 1956. Type: *Lomariopsis* Fée

Lophosoriaceae Pic. Serm. in Webbia 24: 700. 28 Apr 1970 (to be treated as having priority over *Metaxyaceae*). Type: *Lophosoria* C. Presl

Loxogrammaceae Ching ex Pic. Serm. in Webbia 29: 11. 10 Feb 1975. Type: *Loxogramme* (Blume) C. Presl

Loxsomataceae C. Presl, Gefässbündel Farrn: 31. 1847. Type: *Loxsoma* R. Br. ex A. Cunn.

Lycopodiaceae P. Beauv. ex Mirb. in Lamarck & Mirbel, Hist. Nat. Vég. 4: 293. 21 Nov 1802. Type: *Lycopodium* L.

Lygodiaceae M. Roem., Handb. Allg. Bot. 3: 520. 1840. Type: *Lygodium* Sw., *nom. cons.*

Marattiaceae Bercht. & J. Presl, Přir. Rostlin 1: 272. 1820. Type: *Marattia* Sw.

Marsileaceae Mirb. in Lamarck & Mirbel, Hist. Nat. Vég. 5: 126. 21 Nov 1802. Type: *Marsilea* L.

Matoniaceae C. Presl, Gefässbündel Farrn: 32. 1847. Type: *Matonia* R. Br. ex Wall.

Metaxyaceae Pic. Serm. in Webbia 24: 701. 28 Apr 1970 (*Lophosoriaceae* to be treated as having priority). Type: *Metaxya* C. Presl

Monachosoraceae Ching in Acta Phytotax. Sin. 16(4): 17. Nov 1978. Type: *Monachosorum* Kunze

Negripteridaceae Pic. Serm. in Nuovo Giorn. Bot. Ital., ser. 2, 53: 160. [seors. impr.] 25 Nov 1946. Type: *Negripteris* Pic. Serm.

Nephrolepidaceae Pic. Serm. in Webbia 29: 8. 10 Feb 1975. Type: *Nephrolepis* Schott

Oleandraceae (J. Sm.) Ching ex Pic. Serm. in Webbia 20: 745. 20 Dec 1965. Basionym: trib. *Oleandreae* J. Sm., Ferns Brit. For.: 73.

Jun-Jul 1866. Type: *Oleandra* Cav.

Onocleaceae Pic. Serm. in Webbia 24: 708. 28 Apr 1970. Type: *Onoclea* L.

Ophioglossaceae (R. Br.) C. Agardh, Aphor. Bot.: 113. 19 Jun 1822. Basionym: [unranked] *Ophioglosseae* R. Br., Prodr.: 163. 27 Mar 1810. Type: *Ophioglossum* L.

Osmundaceae Bercht. & J. Presl, Přir. Rostlin 1: 272. 1820. Type: *Osmunda* L.

Parkeriaceae Hook., Exot. Fl. 2: sub t. 147. Mar 1825 (rejected in favour of *Adiantaceae;* to be rejected in favour of *Pteridaceae*). Type: *Parkeria* Hook. [= *Ceratopteris* Brongn.]

Peranemataceae (C. Presl) Ching in Sunyatsenia 5: 208, 246. 30 Oct 1940, *nom. cons.* (conserved against *Peranemataceae* Klebs 1892, nom. illeg. [Algae]; rejected in favour of *Dryopteridaceae*). Basionym: trib. *Peranemateae* C. Presl [Tent. Pterid.] in Abh. Königl. Böhm. Ges. Wiss., ser. 4, 5: 64. [seors. impr.] ante 2 Dec 1836. Type: *Peranema* D. Don

Plagiogyriaceae Bower in Ann. Bot. (Oxford) 40: 484. Apr 1926. Type: *Plagiogyria* (Kunze) Mett.

Platyceriaceae (B. K. Nayar) Ching in Acta Phytotax. Sin. 16(3): 18. Aug 1978. Basionym: subfam. *Platycerioideae* B. K. Nayar in Taxon 19: 233. 29 Apr 1970. Type: *Platycerium* Desv.

Platyzomataceae Nakai in Bull. Natl. Sci. Mus. Tokyo 29: 4. 25 Dec 1950. Type: *Platyzoma* R. Br.

Pleurosoriopsidaceae Kurita & Ikebe ex Ching in Acta Phytotax. Sin. 16(4): 17. Nov 1978. Type: *Pleurosoriopsis* Fomin

Polypodiaceae Bercht. & J. Presl, Přir. Rostlin 1: 272. 1820. Type: *Polypodium* L.

Psilotaceae Kanitz, Növényrendsz. Áttekint.: 43. 1887. Type: *Psilotum* Sw.

Pteridaceae Rchb., Handb. Nat. Pfl.-Syst.: 138. 1-7 Oct 1837 (to be conserved against *Parkeriaceae*). Type: *Pteris* L.

Pteridiaceae Ching in Acta Phytotax. Sin. 13(1): 96. Jan 1975. Type: *Pteridium* Gled. ex Scop., *nom. cons.*

Salviniaceae T. Lestib., Botanogr. Elém.: 448. Jun 1826. Type: *Salvinia* Ség.

Schizaeaceae Kaulf., Wesen Farrenkr.: [119]. Jul-Sep 1827. Type: *Schizaea* Sm., *nom. cons.*

Selaginellaceae Willk., Anleit. Stud. Bot. 2: 163. Feb 1854. Type: *Selaginella* P. Beauv., *nom. cons.*

Sinopteridaceae Koidz. in Acta Phytotax. Geobot. 3: 50. 30 Mai 1934. Type: *Sinopteris* C. Chr. & Ching

Stenochlaenaceae Ching in Acta Phytotax. Sin. 16(4): 18. Nov 1978. Type: *Stenochlaena* J. Sm.

Stromatopteridaceae (Nakai) Bierh. in Phytomorphology 18: 263. 15 Dec 1968. Basionym: subfam.

Stromatopteridoideae Nakai in Bull. Natl. Sci. Mus. Tokyo 29: 32. 25 Dec 1950. Type: *Stromatopteris* Mett.

Taenitidaceae (C. Presl) Pic. Serm. in Webbia 29: 1. 10 Feb 1975. Basionym: trib. *Taenitideae* C. Presl [Tent. Pterid.] in Abh. Königl. Böhm. Ges. Wiss., ser. 4, 5: 222. [seors. impr.] ante 2 Dec 1836. Type: *Taenitis* Willd. ex Schkuhr

Tectariaceae Panigrahi in J. Orissa Bot. Soc. 8: 41. 1986. Type: *Tectaria* Cav.

Thelypteridaceae Pic. Serm. in Webbia 24: 709. 28 Apr 1970. Type: *Thelypteris* Schmidel, *nom. cons.*

Thyrsopteridaceae C. Presl, Gefässbündel Farrn: 22, 38. 1847 (rejected in favour of *Dicksoniaceae*). Type: *Thyrsopteris* Kunze

Tmesipteridaceae Nakai, Chosakuronbun Mokuroku [Ord. Fam. Trib. Nov.]: 206. 20 Jul 1943. Type: *Tmesipteris* Bernh.

Trichomanaceae Burmeist., Handb. Naturgesch. 1: 196. 1837. Type: *Trichomanes* L.

Vittariaceae (C. Presl) Ching in Sunyatsenia 5: 210, 232. 30 Oct 1940. Basionym: trib. *Vittarieae* C. Presl [Tent. Pterid.] in Abh. Königl. Böhm. Ges. Wiss., ser. 4, 5: 164. [seors. impr.] ante 2 Dec 1836 Type: *Vittaria* Sm.

Woodsiaceae (Diels) Herter in Revista Sudamer. Bot. 9: 14. Jun 1949 (to be rejected in favour of *Dryopteridaceae*). Basionym: trib. *Woodsieae* Diels in Engler & Prantl, Nat. Pflanzenfam. 1(4): 159. Aug 1899. Type: *Woodsia* R. Br.

Gymnospermae

Amentotaxaceae Kudô & Yamam. in J. Soc. Trop. Agric. 3: 110. Jun 1931. Type: *Amentotaxus* Pilg.

Araucariaceae Henkel & W. Hochst., Syn. Nadelhölz.: xvii, 1. Jan 1865, *nom. cons.* Type: *Araucaria* Juss.

Austrotaxaceae Takht. & Reveal in Phytologia 74: 169. Mar 1993. Type: *Austrotaxus* Compton

Boweniaceae D. W. Stev. in Amer. J. Bot. 68: 1114. 11 Sep 1981. Type: *Bowenia* Hook. f.

Cephalotaxaceae Neger, Nadelhölzer: 23, 30. Oct-Nov 1907, *nom. cons.* Type: *Cephalotaxus* Siebold & Zucc. ex Endl.

Cupressaceae Rich. ex Bartl., Ord. Nat. Pl.: 90, 95. Sep 1830, *nom. cons.* Type: *Cupressus* L.

Cycadaceae Pers., Syn. Pl. 2: 630. Sep 1807, *nom. cons.* Type: *Cycas* L.

Ephedraceae Dumort., Anal. Fam. Pl.: 11, 12. 1829, *nom. cons.* Type: *Ephedra* L.

Ginkgoaceae Engl. in Engler & Prantl, Nat. Pflanzenfam., Nachtr. 1: 19. Jul 1897, *nom. cons.* Type: *Ginkgo* L.

Gnetaceae Lindl. in Edwards's Bot. Reg. 20: sub t. 1686. 1 Jul 1834, *nom. cons.* Type: *Gnetum* L.

Nageiaceae D. Z. Fu in Acta Phytotax. Sin. 30: 522. Dec 1992. Type: *Nageia* Gaertn.

Phyllocladaceae Bessey in Nebraska Univ. Stud. 7: 325. Oct 1907. Type: *Phyllocladus* Rich. ex Mirb., *nom. cons.*

Pinaceae Lindl., Nat. Syst. Bot., ed. 2: 313. Jul 1836, *nom. cons.* Type: *Pinus* L.

Podocarpaceae Endl., Syn. Conif.: 203. Mai-Jun 1847, *nom. cons.* Type: *Podocarpus* L'Hér. ex Pers., *nom. cons.*

Sciadopityaceae Luerss., Grundz. Bot.: 265. Jan 1877. Type: *Sciadopitys* Siebold & Zucc.

Stangeriaceae L. A. S. Johnson in Proc. Linn. Soc. New South Wales, ser. 2, 84: 68. 29 Jul 1959. Type: *Stangeria* T. Moore

Taxaceae Gray, Nat. Arr. Brit. Pl. 2: 222, 226. 1 Nov 1821, *nom. cons.* Type: *Taxus* L.

Taxodiaceae Warm., Haandb. Syst. Bot., ed. 2: 163. Jan-Sep 1884, *nom. cons.* Type: *Taxodium* Rich.

Welwitschiaceae Markgr. in Engler & Prantl, Nat. Pflanzenfam., ed. 2, 13: 419. 1926, *nom. cons.* Type: *Welwitschia* Hook. f., *nom. cons.*

Zamiaceae Horan., Prim. Lin. Syst. Nat.: 45. Jul-Dec 1834. Type: *Zamia* L., *nom. cons.*

Dicotyledones

Acanthaceae Juss., Gen. Pl.: 102. 4 Aug 1789, *nom. cons.* Type: *Acanthus* L.

Aceraceae Juss., Gen. Pl.: 250. 4 Aug 1789, *nom. cons.* (*Sapindaceae* to be treated as having priority). Type: *Acer* L.

Achariaceae Harms in Engler & Prantl, Nat. Pflanzenfam., Nachtr. 1: 256. Oct 1897, *nom. cons.* Type: *Acharia* Thunb.

Achatocarpaceae Heimerl in Engler & Prantl, Nat. Pflanzenfam., ed. 2, 16c: 174. Jan-Apr 1934, *nom. cons.* Type: *Achatocarpus* Triana

Actinidiaceae Hutch., Fam. Fl. Pl. 1: 177. Feb-Mar 1926, *nom. cons.* (conserved against *Saurauiaeae*). Type: *Actinidia* Lindl.

Adoxaceae Trautv., Estestv. Istorija Gub. Kievsk. Učebn. Okr.: 35. 1853, *nom. cons.* (to be conserved against *Sambucaceae* and *Viburnaceae*). Type: *Adoxa* L.

Aegialitidaceae Lincz. in Novosti Sist. Vysš. Rast. 1968: 173. 18 Dec 1968. Type: *Aegialitis* R. Br.

Aegicerataceae Blume, Nov. Pl. Expos.: 17. Aug-Dec 1833. Type: *Aegiceras* Gaertn.

Aextoxicaceae Engl. & Gilg in Engler, Syllabus, ed. 8: 250. Jan-Feb 1920, *nom. cons.* Type: *Aextoxicon* Ruiz & Pav.

Agdestidaceae (Heimerl) Nakai in J. Jap. Bot. 18: 104. Mar 1942. Basionym: trib. *Agdestideae* Heimerl in Engler & Prantl, Nat. Pflanzenfam. 3(1b): 6, 13. Apr 1889. Type: *Agdestis* Moç. & Sessé ex DC.

Aizoaceae F. Rudolphi, Syst. Orb. Veg.: 53. 1830, *nom. cons.* Type: *Aizoon* L.

Akaniaceae Stapf in Bull. Misc. Inform. 1912: 380. 13 Dec 1912, *nom. cons.* Type: *Akania* Hook. f.

Alangiaceae DC., Prodr. 3: 203. Mar 1828, *nom. cons.* Type: *Alangium* Lam., *nom. cons.*

Aldrovandaceae Nakai in J. Jap. Bot. 24: 10. Dec 1949. Type: *Aldrovanda* L.

Alseuosmiaceae Airy Shaw in Kew Bull. 18: 249. 8 Dec 1965. Type: *Alseuosmia* A. Cunn.

Alsinaceae (DC.) Bartl. in Bartling & Wendland, Beitr. Bot. 2: 159. Dec 1825, *nom. cons.* Basionym: trib. *Alsineae* DC., Prodr. 1: 388. Jan 1824. Type: *Alsine* L. [= *Stellaria* L.]

Altingiaceae Lindl., Veg. Kingd.: 253. Jan-Mai 1846, *nom. cons.* Type: *Altingia* Noronha

Alzateaceae S. A. Graham in Ann. Missouri Bot. Gard. 71: 775. 19 Apr 1985. Type: *Alzatea* Ruiz & Pav.

Amaranthaceae Juss., Gen. Pl.: 87. 4 Aug 1789, *nom. cons.* Type: *Amaranthus* L.

Amborellaceae Pichon in Bull. Mus. Natl. Hist. Nat., ser. 2, 20: 384. 25 Oct 1948, *nom. cons.* Type: *Amborella* Baill.

Ambrosiaceae Link, Handbuch 1: 816. Jan-Aug 1829, *nom. cons.* Type: *Ambrosia* L.

Amygdalaceae (Juss.) D. Don, Prodr. Fl. Nepal.: 239. 26 Jan - 1 Feb 1825, *nom. cons.* Basionym: [unranked] *Amygdaleae* Juss., Gen. Pl.: 340. 4 Aug 1789. Type: *Amygdalus* L.

Anacardiaceae Lindl., Intr. Nat. Syst. Bot.: 127. Sep 1830, *nom. cons.* Type: *Anacardium* L.

Ancistrocladaceae Planch. ex Walp. in Ann. Bot. Syst. 2: 175. 15-16 Dec 1851, *nom. cons.* Type: *Ancistrocladus* Wall., *nom. cons.*

Androstachyaceae Airy Shaw in Kew Bull. 18: 250. 8 Dec 1965. Type: *Androstachys* Prain

Anisophylleaceae Ridl., Fl. Malay Penins. 1: 700. 13-22 Jul 1922. Type: *Anisophyllea* R. Br. ex Sabine

Annonaceae Juss., Gen. Pl.: 283. 4 Aug 1789, *nom. cons.* Type: *Annona* L.

Antoniaceae Hutch., Fam. Fl. Pl., ed. 2: 375. Jan-Jul 1959. Type: *Antonia* Pohl

Aphloiaceae Takht. in Bot. Žurn. (Moscow & Leningrad) 70: 1691. 13 Dec 1985. Type: *Aphloia* (DC.) Benn.

Apiaceae Lindl., Nat. Syst. Bot., ed. 2: 21. Jul 1836, *nom. cons.;* nom. alt.: *Umbelliferae.* Type: *Apium* L.

Apocynaceae Juss., Gen. Pl.: 143. 4 Aug 1789, *nom. cons.* Type: *Apocynum* L.

Apodanthaceae (R. Br.) Tiegh. ex Takht., Sist. Magnoliof. [Syst. Magnolioph.]: 42. 31 Mar - 1 Jul 1987. Basionym: trib. *Apodantheae* R. Br. in Trans. Linn. Soc. London 19: 246. 1844. Type:

Apodanthes Poit., non *Apodanthus* Bach. Pyl.

Aptandraceae Miers in Lindley, Veg. Kingd., ed. 3: 447a. Aug 1853. Type: *Aptandra* Miers

Aquifoliaceae Bartl., Ord. Nat. Pl.: 228, 376. Sep 1830, *nom. cons.* Type: *Aquifolium* Hill (*Ilex* L.)

Aquilariaceae R. Br. in Tuckey, Narr. Exped. Zaire: 444. 5 Mar 1818. Type: *Aquilaria* Lam., *nom. cons.*

Araliaceae Juss., Gen. Pl.: 217. 4 Aug 1789, *nom. cons.* Type: *Aralia* L.

Aralidiaceae Philipson & B. C. Stone in Taxon 29: 402. 26 Aug 1980. Type: *Aralidium* Miq.

Argophyllaceae (Engl.) Takht., Sist. Magnoliof. [Syst. Magnolioph.]: 208. 31 Mar - 1 Jul 1987. Basionym: trib. *Argophylleae* Engl. in Engler & Prantl, Nat. Pflanzenfam., ed. 2, 18a: 111, 214. [seors. impr.] 1928. Type: *Argophyllum* J. R. Forst. & G. Forst.

Aristolochiaceae Juss., Gen. Pl.: 72. 4 Aug 1789, *nom. cons.* Type: *Aristolochia* L.

Asclepiadaceae R. Br., Asclepiadeae: 12, 19. 3 Apr 1810, *nom. cons.* Type: *Asclepias* L.

Asteraceae Dumort., Comment. Bot.: 55. Nov-Dec 1822, *nom. cons.;* nom. alt.: *Compositae.* Type: *Aster* L.

Asteranthaceae Knuth in Engler, Pflanzenr. 105: 1. 22 Aug 1939, *nom. cons.* Type: *Asteranthos* Desf.

Asteropeiaceae (Szyszył.) Takht. ex Reveal & Hoogland in Bull. Mus. Natl. Hist. Nat., B, Adansonia 12: 205. 24 Nov 1990. Basionym: trib. *Asteropeieae* Szyszył. in Engler & Prantl, Nat. Pflanzenfam. 3(6): 180, 181. Mai 1893. Type: *Asteropeia* Thouars

Atherospermataceae R. Br. in Flinders, Voy. Terra Austr. 2: 553. 19 Jul 1814. Type: *Atherosperma* Labill.

Aucubaceae J. Agardh, Theoria Syst. Pl.: 303. Apr-Sep 1858. Type: *Aucuba* Thunb.

Austrobaileyaceae (Croizat) Croizat in Cact. Succ. J. (Los Angeles) 15: 64. Mai 1943, *nom. cons.* Basionym: subfam. *Austrobaileyoideae* Croizat in J. Arnold Arbor. 21: 404. 24 Jul 1940. Type: *Austrobaileya* C. T. White

Averrhoaceae Hutch., Fam. Fl. Pl., ed. 2: 356. Jan-Jul 1959. Type: *Averrhoa* L.

Avicenniaceae Endl., Ench. Bot.: 314. 15-21 Aug 1841, *nom. cons.* Type: *Avicennia* L.

Balanitaceae Endl., Ench. Bot.: 547. 15-21 Aug 1841, *nom. cons.* Type: *Balanites* Delile, *nom. cons.*

Balanopaceae Benth. in Bentham & Hooker, Gen. Pl. 3: v, 341. 7 Feb 1880, *nom. cons.* Type: *Balanops* Baill.

Balanophoraceae Rich. in Mém. Mus. Hist. Nat. 8: 429. Nov 1822, *nom. cons.* Type: *Balanophora* J. R. Forst. & G. Forst.

Balsaminaceae A. Rich. in Bory, Dict. Class. Hist. Nat. 2: 173. 31 Dec 1822, *nom. cons.* Type: *Balsamina* Hill [= *Impatiens* L.]

Barbeuiaceae (H. Walter) Nakai in J. Jap. Bot. 18: 105. Mar 1942. Basionym: subtrib. *Barbeuiinae* H. Walter in Engler, Pflanzenr. 39: 28, 63. 1 Jun 1909. Type: *Barbeuia* Thouars

Barbeyaceae Rendle in Oliver, Fl. Trop. Afr. 6(2): 14. Mar 1916, *nom. cons.* Type: *Barbeya* Schweinf.

Barclayaceae (Endl.) H. L. Li in Amer. Midl. Naturalist 54: 40. 27 Aug 1955. Basionym: trib. *Barclayeae* Endl., Gen. Pl.: 900. Nov 1839. Type: *Barclaya* Wall.

Barringtoniaceae F. Rudolphi, Syst. Orb. Veg.: 56. 1830, *nom. cons.* Type: *Barringtonia* J. R. Forst. & G. Forst., *nom. cons.*

Basellaceae Moq., Chenop. Monogr. Enum.: x. Mai 1840, *nom. cons.* Type: *Basella* L.

Bataceae Mart. ex Meisn., Pl. Vasc. Gen., Tab. Diagn.: 349; Comm.: 260. 13-15 Feb 1842, *nom. cons.* Type: *Batis* P. Browne

Baueraceae Lindl., Intr. Nat. Syst. Bot.: 50. Sep 1830. Type: *Bauera* Banks ex Andrews

Begoniaceae C. Agardh, Aphor. Bot.: 200. 13 Jun 1824, *nom. cons.* Type: *Begonia* L.

Berberidaceae Juss., Gen. Pl.: 286. 4 Aug 1789, *nom. cons.* Type: *Berberis* L.

Berberidopsidaceae (Veldkamp) Takht. in Bot. Žurn. (Moscow & Leningrad) 70: 1691. 13 Dec 1985. Basionym: trib. *Berberidopsideae* Veldkamp in Blumea 30: 24. 12 Nov 1984. Type: *Berberidopsis* Hook. f.

Betulaceae Gray, Nat. Arr. Brit. Pl. 2: 222, 243. 1 Nov 1821, *nom. cons.* (conserved against *Corylaceae;* to be conserved against *Carpinaceae*). Type: *Betula* L.

Biebersteiniaceae Endl., Ench. Bot.: 618. 15-21 Aug 1841. Type: *Biebersteinia* Stephan

Bignoniaceae Juss., Gen. Pl.: 137. 4 Aug 1789, *nom. cons.* Type: *Bignonia* L.

Bischofiaceae (Müll. Arg.) Airy Shaw in Kew Bull. 18: 252. 8 Dec 1965. Basionym: subtrib. *Bischofiinae* Müll. Arg. in Linnaea 34: 64. Mar 1865. Type: *Bischofia* Blume

Bixaceae Link, Handbuch 2: 371. Jan-Aug 1831, *nom. cons.* Type: *Bixa* L.

Blepharocaryaceae Airy Shaw in Kew Bull. 18: 254. 8 Dec 1965. Type: *Blepharocarya* F. Muell.

Boerlagellaceae H. J. Lam in Bull. Jard. Bot. Buitenzorg, ser. 3, 7: 250. Feb 1925. Type: *Boerlagella* Cogn.

Bombacaceae Kunth, Malvac., Büttner., Tiliac.: 5. 20 Apr 1822, *nom. cons.* Type: *Bombax* L.

Bonnetiaceae (Bartl.) L. Beauvis. ex Nakai in Bull. Tokyo Sci. Mus. 22: 25. 1948. Basionym: [unranked] *Bonnetiea* Bartl., Ord. Nat. Pl.: 336. Sep 1830. Type: *Bonnetia* Mart., *nom. cons.*

Boraginaceae Juss., Gen. Pl.: 128. 4 Aug 1789, *nom. cons.* Type: *Borago* L.

Brassicaceae Burnett, Outl. Bot.: 854, 1123. Jun 1835, *nom. cons.;* nom. alt.: *Cruciferae.* Type: *Brassica* L.

Bretschneideraceae Engl. & Gilg in Engler, Syllabus, ed. 9 & 10: 218. Nov-Dec 1924, *nom. cons.* Type: *Bretschneidera* Hemsl.

Brexiaceae Loudon, Hort. Brit.: 524. 1-7 Sep 1830. Type: *Brexia* Noronha ex Thouars, *nom. cons.*

Brunelliaceae Engl. in Engler & Prantl, Nat. Pflanzenfam., Nachtr. 1: 182. Aug 1897, *nom. cons.* Type: *Brunellia* Ruiz & Pav.

Bruniaceae R. Br. ex DC., Prodr. 2: 43. Nov 1825, *nom. cons.* Type: *Brunia* Lam., *nom. cons.*

Brunoniaceae Dumort., Anal. Fam. Pl.: 19, 21. 1829, *nom. cons.* Type: *Brunonia* Sm. ex R. Br.

Buddlejaceae K. Wilh., Samenpflanzen: 90. Oct 1910, *nom. cons.* Type: *Buddleja* L.

Burseraceae Kunth in Ann. Sci. Nat. (Paris) 2: 346. Jun 1824, *nom. cons.* Type: *Bursera* Jacq. ex L., *nom. cons.*

Buxaceae Dumort., Comment. Bot.: 54. Nov-Dec 1822, *nom. cons.* Type: *Buxus* L.

Byblidaceae (Engl. & Gilg) Domin in Acta Bot. Bohem. 1: 3. 1922, *nom. cons.* Basionym: subfam. *Byblidoideae* Engl. & Gilg in Engler, Syllabus, ed. 7: 329. Oct 1912 - Mar 1913. Type: *Byblis* Salisb.

Byttneriaceae R. Br. in Flinders, Voy. Terra Austr. 2: 540. 19 Jul 1814, *nom. cons.* (rejected in favour of *Sterculiaceae*). Type: *Byttneria* Loefl., *nom. cons.*

Cabombaceae A. Rich., Nouv. Elém. Bot., ed. 4: 420. 1828, *nom. cons.* Type: *Cabomba* Aubl.

Cactaceae Juss., Gen. Pl.: 310. 4 Aug 1789, *nom. cons.* Type: *Cactus* L., *nom. rej. (Mammillaria* Haw., *nom. cons.)*

Caesalpiniaceae R. Br. in Flinders, Voy. Terra Austr. 2: 551. 19 Jul 1814, *nom. cons.* Type: *Caesalpinia* L.

Calligonaceae Khalk. in Dokl. Akad. Nauk. Uzbeksk. S.S.R. 1985(11): 45. 1985. Type: *Calligonum* L.

Callitrichaceae Link, Enum. Hort. Berol. Alt. 1: 7. 16 Mar - 30 Jun 1821, *nom. cons.* Type: *Calliriche* L.

Calycanthaceae Lindl. in Bot. Reg. 5: sub t. 404. 1 Oct 1819, *nom. cons.* Type: *Calycanthus* L., *nom. cons.*

Calyceraceae R. Br. ex Rich. in Mém. Mus. Hist. Nat. 6: 74. Nov 1820, *nom. cons.* Type: *Calycera* Cav., *orth. cons. prop.*

Campanulaceae Juss., Gen. Pl.: 163. 4 Aug 1789, *nom. cons.* Type: *Campanula* L.

Canellaceae Mart., Nov. Gen. Sp. Pl. 3: 170. Sep 1832, *nom. cons.* Type: *Canella* P. Browne, *nom. cons.*

Cannabaceae Endl., Gen. Pl.: 286. Oct 1837, *nom. cons.* Type: *Cannabis* L.

Canotiaceae Airy Shaw in Kew Bull. 18: 255. 8 Dec 1965. Type: *Canotia* Torr.

Capparaceae Juss., Gen. Pl.: 242. 4 Aug 1789, *nom. cons.* Type: *Capparis* L.

Caprifoliaceae Juss., Gen. Pl.: 210. 4 Aug 1789, *nom. cons.* Type: *Caprifolium* Hill (*Lonicera* L.)

Cardiopteridaceae Blume, Rumhia 3: 205. Jun 1847, *nom. cons.* Type: *Cardiopteris* Wall. ex Royle

Caricaceae Dumort., Anal. Fam. Pl.: 37, 42. 1829, *nom. cons.* Type: *Carica* L.

Carlemanniaceae Airy Shaw in Kew Bull. 19: 511. 26 Jul 1965. Type: *Carlemannia* Benth.

Carpinaceae Vest, Anleit. Stud. Bot.: 265, 280. 1818 (to be rejected in favour of *Betulaceae*). Type: *Carpinus* L.

Carpodetaceae Fenzl in Denkschr. Königl.-Baier. Bot. Ges. Regensburg 3: 155. 28 Nov 1841. Type: *Carpodetus* J. R. Forst. & G. Forst.

Caryocaraceae Szyszył. in Engler & Prantl, Nat. Pflanzenfam. 3(6): 153. Mai 1893, *nom. cons.* Type: *Caryocar* L.

Caryophyllaceae Juss., Gen. Pl.: 299. 4 Aug 1789, *nom. cons.* Type: *Caryophyllus* Hill non L. (*Dianthus* L.)

Cassythaceae Bartl. ex Lindl., Nix. Pl.: 15. 17 Sep 1833, *nom. cons.* Type: *Cassytha* L.

Casuarinaceae R. Br. in Flinders, Voy. Terra Austr. 2: 571. 19 Jul 1814, *nom. cons.* Type: *Casuarina* L.

Cecropiaceae C. C. Berg in Taxon 27: 39. 23 Mar 1978. Type: *Cecropia* Loefl., *nom. cons.*

Celastraceae R. Br. in Flinders, Voy. Terra Austr. 2: 554. 19 Jul 1814, *nom. cons.* (conserved against *Hippocrateaceae*). Type: *Celastrus* L.

Celtidaceae Link, Handbuch 2: 441. Jan-Aug 1831. Type: *Celtis* L.

Cephalotaceae Dumort., Anal. Fam. Pl.: 59, 61. 1829, *nom. cons.* Type: *Cephalotus* Labill., *nom. cons.*

Ceratophyllaceae Gray, Nat. Arr. Brit. Pl. 2: 395, 554. 1 Nov 1821, *nom. cons.* Type: *Ceratophyllum* L.

Cercidiphyllaceae Engl., Syllabus, ed. 6: 132. Jun-Dec 1909, *nom. cons.* Type: *Cercidiphyllum* Siebold & Zucc.

Chailletiaceae – see *Dichapetalaceae*

Chelidoniaceae Martinov, Tekhno-Bot. Slovar: 124. 1820. Type: *Chelidonium* L.

Chenopodiaceae Vent., Tabl. Règne Vég. 2: 253. 5 Mai 1799, *nom. cons.* Type: *Chenopodium* L.

Chloanthaceae Hutch., Fam. Fl. Pl., ed. 2: 396. Jan-Jul 1959. Type: *Chloanthes* R. Br.

Chloranthaceae R. Br. ex Lindl., Coll. Bot.: sub t. 17. 21 Oct 1821, *nom. cons.* Type: *Chloranthus* Sw.

Chrysobalanaceae R. Br. in Tuckey, Narr. Exped. Zaire: 433. 5 Mar 1818, *nom. cons.* Type: *Chrysobalanus* L.

Cichoriaceae Juss., Gen. Pl.: 168. 4 Aug 1789, *nom. cons.* Type: *Cichorium* L.

Circaeasteraceae Hutch., Fam. Fl. Pl. 1: 98. Feb-Mar 1926, *nom. cons.* Type: *Circaeaster* Maxim.

Cistaceae Juss., Gen. Pl.: 294. 4 Aug 1789, *nom. cons.* Type: *Cistus* L.

Cleomaceae Horan., Prim. Lin. Syst. Nat.: 92. Jul-Dec 1834. Type: *Cleome* L.

Clethraceae Klotzsch in Linnaea 24: 12. Mai 1851, *nom. cons.* Type: *Clethra* L.

Clusiaceae Lindl., Nat. Syst. Bot., ed. 2: 74. Jul 1836, *nom. cons.;* nom. alt.: *Guttiferae.* Type: *Clusia* L.

Cneoraceae Link, Handbuch 2: 440. Jan-Aug 1831, *nom. cons.* Type: *Cneorum* L.

Cobaeaceae D. Don in Edinburgh Philos. J. 10: 111. Jan 1824. Type: *Cobaea* Cav.

Cochlospermaceae Planch. in London J. Bot. 6: 305. Jun-Jul 1847, *nom. cons.* Type: *Cochlospermum* Kunth, *nom. cons.*

Columelliaceae D. Don in Edinburgh New Philos. J. 6: 46, 49. 1828, *nom. cons.* Type: *Columellia* Ruiz & Pav., *nom. cons.*

Combretaceae R. Br., Prodr.: 351. 27 Mar 1810, *nom. cons.* Type: *Combretum* Loefl., *nom. cons.*

Compositae Giseke, Prael. Ord. Nat. Pl.: 538. Apr 1792, *nom. cons.;* nom. alt.: *Asteraceae.* Type: *Aster* L.

Connaraceae R. Br. in Tuckey, Narr. Exped. Zaire: 431. 5 Mar 1818, *nom. cons.* Type: *Connarus* L.

Convolvulaceae Juss., Gen. Pl.: 132. 4 Aug 1789, *nom. cons.* Type: *Convolvulus* L.

Cordiaceae R. Br. ex Dumort., Anal. Fam. Pl.: 20, 25. 1829, *nom. cons.* Type: *Cordia* L.

Coriariaceae DC., Prodr. 1: 739. Jan 1824, *nom. cons.* Type: *Coriaria* L.

Coridaceae J. Agardh, Theoria Syst. Pl.: 332. Apr-Sep 1858. Type: *Coris* L.

Cornaceae (Dumort.) Dumort., Anal. Fam. Pl.: 33, 34. 1829, *nom. cons.* (to be treated as having priority over *Nyssaceae*). Basionym: trib. *Corneae* Dumort., Fl. Belg.: 83. 1827-1830. Type: *Cornus* L.

Corylaceae Mirb., Elém. Physiol. Vég. Bot. 2: 906. 1815, *nom. cons.*

(rejected in favour of *Betulaceae*). Type: *Corylus* L.

Corynocarpaceae Engl. in Engler & Prantl, Nat. Pflanzenfam., Nachtr. 1: 215. Oct 1897, *nom. cons.* Type: *Corynocarpus* J. R. Forst. & G. Forst.

Crassulaceae DC. in Lamarck & Candolle, Fl. Franç., ed. 3, 4(1): 382. 17 Sep 1805, *nom. cons.* Type: *Crassula* L.

Crossosomataceae Engl. in Engler & Prantl, Nat. Pflanzenfam., Nachtr. 1: 185. Aug 1897, *nom. cons.* Type: *Crossosoma* Nutt.

Cruciferae Juss., Gen. Pl.: 237. 4 Aug 1789, *nom. cons.;* nom. alt.: *Brassicaceae.* Type: *Brassica* L.

Crypteroniaceae A. DC. in Candolle, Prodr. 16(2): 677. Jul 1868, *nom. cons.* Type: *Crypteronia* Blume

Ctenolophaceae (H. Winkl.) Exell & Mendonça, Consp. Fl. Angol. 1: 248, 392. 20 Aug 1951. Basionym: subfam. *Ctenolophoideae* H. Winkl. in Engler & Prantl, Nat. Pflanzenfam., ed. 2, 19a: 106, 122. Jan-Aug 1931. Type: *Ctenolophon* Oliv.

Cucurbitaceae Juss., Gen. Pl.: 393. 4 Aug 1789, *nom. cons.* Type: *Cucurbita* L.

Cunoniaceae R. Br. in Flinders, Voy. Terra Austr. 2: 548. 19 Jul 1814, *nom. cons.* Type: *Cunonia* L., *nom. cons.*

Curtisiaceae (Harms) Takht., Sist. Magnoliof. [Syst. Magnolioph.]: 214. 31 Mar - 1 Jul 1987. Basionym: subfam. *Curtisioideae* Harms in Engler & Prantl, Nat. Pflanzenfam. 3(8): 255, 262. Jul 1898. Type: *Curtisia* Aiton, *nom. cons.*

Cuscutaceae (Dumort.) Dumort., Anal. Fam. Pl.: 20, 25. 1829, *nom. cons.* Basionym: trib. *Cuscuteae* Dumort., Fl. Belg.: 50. 1827-1830. Type: *Cuscuta* L.

Cyclocheilaceae Marais in Kew Bull. 35: 805. 28 Apr 1981. Type: *Cyclocheilon* Oliv.

Cynomoriaceae (C. Agardh) Lindl., Nix. Pl.: 23. 17 Sep 1833, *nom. cons.* Basionym: [unranked] *Cynomorieae* C. Agardh, Aphor. Bot.: 203. 13 Jun 1824. Type: *Cynomorium* L.

Cyphiaceae A. DC. in Candolle, Prodr. 7(2): 497. Dec 1839. Type: *Cyphia* P. J. Bergius

Cyphocarpaceae Miers in London J. Bot. 7: 61. Jan-Feb 1848. Type: *Cyphocarpus* Miers

Cyrillaceae Endl., Ench. Bot.: 578. 15-21 Aug 1841, *nom. cons.* Type: *Cyrilla* Garden ex L.

Cytinaceae (Brongn.) A. Rich. in Bory, Dict. Class. Hist. Nat. 5: 301. 15 Mai 1824 (to be rejected in favour of *Rafflesiaceae*). Basionym: trib. *Cytineae* Brongn. in Ann. Sci. Nat. (Paris) 1: 39. Jan-Apr 1824. Type: *Cytinus* L., *nom. cons.*

Dactylanthaceae (Engl.) Takht., Sist. Magnoliof. [Syst. Magnolioph.]: 43. 31 Mar - 1 Jul 1987. Basionym: subfam. *Dactylanthoideae* Engl. in Engler & Prantl, Nat. Pflanzenfam. 3(1): 250, 252.

Aug 1889. Type: *Dactylanthus* Hook. f.

Daphniphyllaceae Müll. Arg. in Candolle, Prodr. 16(1): 1. Nov 1869, *nom. cons.* Type: *Daphniphyllum* Blume

Datiscaceae R. Br. ex Lindl., Intr. Nat. Syst. Bot.: 109. Sep 1830, *nom. cons.* Type: *Datisca* L.

Davidiaceae (Harms) H. L. Li in Lloydia 17: 330. 18 Mar 1955. Basionym: subfam. *Davidioideae* Harms in Engler & Prantl, Nat. Pflanzenfam. 3(8): 255, 259. Jul 1898. Type: *Davidia* Baill.

Davidsoniaceae Bange in Blumea 7: 294. 1 Sep 1952. Type: *Davidsonia* F. Muell.

Degeneriaceae I. W. Bailey & A. C. Sm. in J. Arnold Arbor. 23: 357. 15 Jul 1942, *nom. cons.* Type: *Degeneria* I. W. Bailey & A. C. Sm.

Desfontainiaceae Endl., Ench. Bot.: 336. 15-21 Aug 1841, *nom. cons.* Type: *Desfontainia* Ruiz & Pav.

Dialypetalanthaceae Rizzini & Occhioni in Lilloa 17: 253. 30 Dec 1948, *nom. cons.* Type: *Dialypetalanthus* Kuhlm.

Diapensiaceae (Link) Lindl., Nat. Syst. Bot., ed. 2: 233. Jul 1836, *nom. cons.* Basionym: subfam. *Diapensioideae* Link, Handbuch 1: 595. Jan-Aug 1829. Type: *Diapensia* L.

Dichapetalaceae Baill. in Martius, Fl. Bras. 12(1): 365. 1 Apr 1886, *nom. cons.* Type: *Dichapetalum* Thouars

Dichondraceae Dumort., Anal. Fam. Pl.: 20, 24. 1829, *nom. cons.* Type: *Dichondra* J. R. Forst. & G. Forst.

Diclidantheraceae J. Agardh, Theoria Syst. Pl.: 195. Apr-Sep 1858, *nom. cons.* Type: *Diclidanthera* Mart.

Dicrastylidaceae – see *Chloanthaceae*

Didiereaceae Drake in Bull. Mus. Hist. Nat. (Paris) 9: 36. 17 Feb 1903, *nom. cons.* Type: *Didierea* Baill.

Didymelaceae Leandri in Ann. Sci. Nat., Bot., ser. 10, 19: 316. Mai 1937. Type: *Didymeles* Thouars

Diegodendraceae Capuron in Adansonia, ser. 2, 3: 392. Oct-Dec 1964. Type: *Diegodendron* Capuron

Dilleniaceae Salisb., Parad. Lond. 2: sub t. 73. Jun 1807, *nom. cons.* Type: *Dillenia* L.

Dionaeaceae Raf., Fl. Tellur. 3: 35. Nov-Dec 1837. Type: *Dionaea* Sol. ex J. Ellis

Dioncophyllaceae (Gilg) Airy Shaw in Kew Bull. 1951: 333. 26 Jan 1952, *nom. cons.* Basionym: subtrib. *Dioncophyllinae* Gilg in Engler & Prantl, Nat. Pflanzenfam., ed. 2, 21: 390, 420. Dec 1925. Type: *Dioncophyllum* Baill.

Dipentodontaceae Merr. in Brittonia 4: 69. 16 Dec 1941, *nom. cons.* Type: *Dipentodon* Dunn

Dipsacaceae Juss., Gen. Pl.: 194. 4 Aug 1789, *nom. cons.* Type: *Dipsacus* L.

Dipterocarpaceae Blume, Bijdr. Fl. Ned. Ind.: 222. 20 Sep - 7 Dec 1825, *nom. cons.* Type: *Dipterocarpus* C. F. Gaertn.

Dirachmaceae Hutch., Fam. Fl. Pl., ed. 2: 248. Jan-Jul 1959. Type: *Dirachma* Schweinf. ex Balf. f.

Dodonaeaceae Link, Handbuch 2: 441. Jan-Aug 1831, *nom. cons.* Type: *Dodonaea* Mill.

Donatiaceae Takht. ex Dostál, Bot. Nomenkl.: 204. 1957, *nom. cons.* Type: *Donatia* J. R. Forst. & G. Forst., *nom. cons.*

Droseraceae Salisb., Parad. Lond. 2: sub t. 95. 1808, *nom. cons.* Type: *Drosera* L.

Drosophyllaceae Chrtek & al. in Preslia 61: 122. 10 Apr 1989. Type: *Drosophyllum* Link

Duabangaceae Takht., Florist. Reg. World: 332. 4 Sep 1986. Type: *Duabanga* Buch.-Ham.

Duckeodendraceae Kuhlm. in Arq. Serv. Florest. 3: 7. 1950. Type: *Duckeodendron* Kuhlm.

Dulongiaceae, nom. illeg. – see *Phyllonomaceae*

Dysphaniaceae Pax in Bot. Jahrb. Syst. 61: 230. 15 Jun 1927, *nom. cons.* Type: *Dysphania* R. Br.

Ebenaceae Gürke in Engler & Prantl, Nat. Pflanzenfam. 4(1): 153. Dec 1891, *nom. cons.* Type: *Ebenus* Kuntze, non L. (*Maba* J. R. Forst. & G. Forst.)

Ehretiaceae Mart. ex Lindl., Intr. Nat. Syst. Bot.: 242. Sep 1830, *nom. cons.* Type: *Ehretia* P. Browne

Elaeagnaceae Juss., Gen. Pl.: 74. 4 Aug 1789, *nom. cons.* Type: *Elaeagnus* L.

Elaeocarpaceae Juss. ex DC., Prodr. 1: 519. Jan 1824, *nom. cons.* Type: *Elaeocarpus* L.

Elatinaceae Dumort., Anal. Fam. Pl.: 44, 49. 1829, *nom. cons.* Type: *Elatine* L.

Ellisiophyllaceae Honda in Honda & Sakisaka, Taikou Nippon Shokubutsu Bunraigaku [Syst. Pl. Japon.]: 372. 1930. Type: *Ellisiophyllum* Maxim.

Emblingiaceae (Pax) Airy Shaw in Kew Bull. 18: 257. 8 Dec 1965. Basionym: subfam. *Emblingioideae* Pax in Engler & Prantl, Nat. Pflanzenfam. 3(2): 220, 235. Mar 1891. Type: *Emblingia* F. Muell.

Empetraceae Gray, Nat. Arr. Brit. Pl. 2: 732. 1 Nov 1821, *nom. cons.* Type: *Empetrum* L.

Epacridaceae R. Br., Prodr.: 535. 27 Mar 1810, *nom. cons.* Type: *Epacris* Cav., *nom. cons.*

Eremolepidaceae Tiegh. ex Nakai in Bull. Natl. Sci. Mus. Tokyo 31: 48. Mar 1952. Type: *Eremolepis* Griseb.

Eremosynaceae Dandy in Hutchinson, Fam. Fl. Pl., ed. 2: 460. Jan-Jul 1959. Type: *Eremosyne* Endl.

Ericaceae Juss., Gen. Pl.: 159. 4 Aug 1789, *nom. cons.* Type: *Erica* L.

Erythropalaceae (Hassk.) Sleumer in Engler & Prantl, Nat. Pflanzenfam., ed. 2, 20b: 401. Jan-Feb 1942, *nom. cons.* Basionym: subfam. *Erythropaloideae* Hassk., Pl.

Jav. Rar.: 193. Jan-Feb 1848. Type: *Erythropalum* Blume

Erythroxylaceae Kunth in Humboldt & al., Nov. Gen. Sp. 5, ed. 4°: 175 [ed. f°: 135]. 25 Feb 1822, *nom. cons.* Type: *Erythroxylum* P. Browne

Escalloniaceae R. Br. ex Dumort., Anal. Fam. Pl.: 35, 37. 1829, *nom. cons.* Type: *Escallonia* Mutis ex L. f.

Eschscholziaceae Ser., Fl. Jard. 2: 106. Apr 1847. Type: *Eschscholzia* Cham.

Eucommiaceae Engl., Syllabus, ed. 6: 145. Jun-Dec 1909, *nom. cons.* Type: *Eucommia* Oliv.

Eucryphiaceae Endl., Ench. Bot.: 528. 15-21 Aug 1841, *nom. cons.* Type: *Eucryphia* Cav.

Euphorbiaceae Juss., Gen. Pl.: 384. 4 Aug 1789, *nom. cons.* Type: *Euphorbia* L.

Euphroniaceae Marc.-Berti in Pittiera 18: 16. Nov 1989. Type: *Euphronia* Mart. & Zucc.

Eupomatiaceae Endl., Ench. Bot.: 425. 15-21 Aug 1841, *nom. cons.* Type: *Eupomatia* R. Br.

Eupteleaceae K. Wilh., Samenpflanzen: 17. Oct 1910, *nom. cons.* Type: *Euptelea* Siebold & Zucc.

Euryalaceae J. Agardh, Theoria Syst. Pl.: 51. Apr-Sep 1858. Type: *Euryale* Salisb.

Fabaceae Lindl., Nat. Syst. Bot., ed. 2: 148. Jul 1836, *nom. cons.;* nom. alt.: *Leguminosae, Papilionaceae.* Type: *Faba* Hill [= *Vicia* L.]

Fagaceae Dumort., Anal. Fam. Pl.: 11, 12. 1829, *nom. cons.* Type: *Fagus* L.

Flacourtiaceae Rich. ex DC., Prodr. 1: 255. Jan 1824, *nom. cons.* (conserved against *Samydaceae*). Type: *Flacourtia* Comm. ex L'Hér.

Flindersiaceae (Engl.) C. T. White ex Airy Shaw in Kew Bull. 18: 257. 8 Dec 1965. Basionym: subfam. *Flindersioideae* Engl. in Engler & Prantl, Nat. Pflanzenfam. 3(4): 111, 170. Mar 1896. Type: *Flindersia* R. Br.

Foetidiaceae (Nied.) Airy Shaw in Kew Bull. 18: 258. 8 Dec 1956. Basionym: subfam. *Foetidioideae* Nied. in Engler & Prantl, Nat. Pflanzenfam. 3(7): 29 Jun 1892. Type: *Foetidia* Comm. ex Lam.

Fouquieriaceae DC., Prodr. 3: 349. Mar 1828, *nom. cons.* Type: *Fouquieria* Kunth

Francoaceae A. Juss. in Ann. Sci. Nat. (Paris) 25: 9. Jan 1832. Type: *Francoa* Cav.

Frankeniaceae A. St.-Hil. ex Gray, Nat. Arr. Brit. Pl. 2: 623, 633. 1 Nov 1821, *nom. cons.* Type: *Frankenia* L.

Fumariaceae DC., Syst. Nat. 2: 65, 105. Mai 1821, *nom. cons.* Type: *Fumaria* L.

Garryaceae Lindl. in Edwards's Bot. Reg. 20: sub t. 1686. 1 Jul 1834, *nom. cons.* Type: *Garrya* Douglas ex Lindl.

Geissolomataceae Endl., Ench. Bot.: 214. 15-21 Aug 1841, *nom. cons.* Type: *Geissoloma* Lindl. ex Kunth

Gentianaceae Juss., Gen. Pl.: 141. 4 Aug 1789, *nom. cons.* Type: *Gentiana* L.

Geraniaceae Juss., Gen. Pl.: 268. 4 Aug 1789, *nom. cons.* Type: *Geranium* L.

Gesneriaceae Dumort., Comment. Bot.: 57. Nov-Dec 1822, *nom. cons.* Type: *Gesneria* L.

Gisekiaceae (Endl.) Nakai in J. Jap. Bot. 18: 102. Mar 1942. Basionym: trib. *Gisekieae* Endl., Gen. Pl.: 977. 1-14 Feb 1840. Type: *Gisekia* L.

Glaucidiaceae Tamura in Bot. Mag. (Tokyo) 85: 40. Mar 1972. Type: *Glaucidium* Siebold & Zucc.

Globulariaceae DC. in Lamarck & Candolle, Fl. Franç., ed. 3, 3: 427. 17 Sep 1805, *nom. cons.* Type: *Globularia* L.

Goetzeaceae Miers ex Airy Shaw in Kew Bull. 18: 259. 8 Dec 1965. Type: *Goetzea* Wydler, *nom. cons.*

Gomortegaceae Reiche in Ber. Deutsch. Bot. Ges. 14: 232. 19 Aug 1896, *nom. cons.* Type: *Gomortega* Ruiz & Pav.

Gonystylaceae Gilg in Engler & Prantl, Nat. Pflanzenfam., Nachtr. 1: 231. Oct 1897, *nom. cons.* Type: *Gonystylus* Teijsm. & Binn.

Goodeniaceae R. Br., Prodr.: 573. 27 Mar 1810, *nom. cons.* Type: *Goodenia* Sm.

Goupiaceae Miers in Ann. Mag. Nat. Hist., ser. 3, 9: 292. Apr 1862. Type: *Goupia* Aubl.

Greyiaceae Hutch., Fam. Fl. Pl. 1: 202. Feb-Mar 1926, *nom. cons.* Type: *Greyia* Hook. & Harv.

Griseliniaceae (Wangerin) Takht., Sist. Magnoliof. [Syst. Magnolioph.]: 209. 31 Mar - 1 Jul 1987. Basionym: trib. *Griselinieae* Wangerin in Engler, Pflanzenr. 41: 94. 12 Apr 1910. Type: *Griselinia* J. R. Forst. & G. Forst.

Grossulariaceae DC. in Lamarck & Candolle, Fl. Franç., ed. 3, 4(2): 405. 17 Sep 1805, *nom. cons.* Type: *Grossularia* Mill.

Grubbiaceae Endl., Gen. Pl.: xiv. Jan 1839, *nom. cons.* Type: *Grubbia* P. J. Bergius

Gunneraceae Meisn., Pl. Vasc. Gen., Tab. Diagn.: 345, 346; Comm.: 257. 13-15 Feb 1842, *nom. cons.* Type: *Gunnera* L.

Guttiferae Juss., Gen. Pl.: 255. 4 Aug 1789, *nom. cons.,* nom. alt.: *Clusiaceae.* Type: *Clusia* L.

Gyrocarpaceae Dumort., Anal. Fam. Pl.: 13, 14. 1829. Type: *Gyrocarpus* Jacq.

Gyrostemonaceae Endl., Ench. Bot.: 509. 15-21 Aug 1841, *nom. cons.* Type: *Gyrostemon* Desf.

Halophytaceae A. Soriano in Bol. Soc. Argent. Bot. 23: 161. 29 Aug 1984. Type: *Halophytum* Speg.

Haloragaceae R. Br. in Flinders, Voy. Terra Austr. 2: 549. 19 Jul 1814, *nom. cons.* Type: *Haloragis* J. R. Forst. & G. Forst.

Hamamelidaceae R. Br. in Abel, Narr. Journey China: 374. Aug 1818, *nom. cons.* Type: *Hamamelis* L.

Hectorellaceae Philipson & Skipw. in Trans. Roy. Soc. New Zealand, Bot. 1: 31. 20 Sep 1961. Type: *Hectorella* Hook. f.

Heliamphoraceae Chrtek & al. in Preslia 64: 8. 10 Feb 1992. Type: *Heliamphora* Benth.

Heliotropiaceae Schrad. in Commentat. Soc. Regiae Sci. Gott. Recent. 4: 192. Dec 1819, *nom. cons.* Type: *Heliotropium* L.

Helleboraceae Vest, Anleit. Stud. Bot.: 264, 274. 1818. Type: *Helleborus* L.

Helosaceae (Schott & Endl.) Tiegh. ex Reveal & Hoogland in Bull. Mus. Natl. Hist. Nat., B, Adansonia 12: 205. 24 Nov 1990. Basionym: trib. *Heloseae* Schott & Endl., Melet. Bot.: 11. 1832. Type: *Helosis* Rich., *nom. cons.*

Helwingiaceae Decne. in Ann. Sci. Nat., Bot., ser. 2, 6: 69. Aug 1836. Type: *Helwingia* Willd., *nom. cons.*

Henriqueziaceae Bremek. in Acta Bot. Neerl. 6: 371. Aug 1957. Type: *Henriquezia* Spruce ex Benth.

Hernandiaceae Blume, Bijdr. Fl. Ned. Ind.: 550. 24 Jan 1826, *nom. cons.* Type: *Hernandia* L.

Heteropyxidaceae Engl. & Gilg in Engler, Syllabus, ed. 8: 281. Jan-Feb 1920, *nom. cons.* Type: *Heteropyxis* Harv., *nom. cons.*

Himantandraceae Diels in Bot. Jahrb. Syst. 55: 126. 27 Nov 1917, *nom. cons.* Type: *Himantandra* F. Muell. ex Diels [= *Galbulimima* F. M. Bailey]

Hippocastanaceae DC., Prodr. 1: 597. Jan 1824, *nom. cons.* Type: *Hippocastanum* Hill (*Aesculus* L.)

Hippocrateaceae Juss. in Ann. Mus. Natl. Hist. Nat. 18: 486. Jul-Aug 1811, *nom. cons.* (rejected in favour of *Celastraceae*). Type: *Hippocratea* L.

Hippuridaceae Link, Enum. Hort. Berol. Alt. 1: 5. 16 Mar - 30 Jun 1821, *nom. cons.* Type: *Hippuris* L.

Hoplestigmataceae Gilg in Engler, Syllabus, ed. 9 & 10: 322. Nov-Dec 1924, *nom. cons.* Type: *Hoplestigma* Pierre

Hortoniaceae (Perkins & Gilg) A. C. Sm. in J. Indian Bot. Soc. 50A: 220. Dec 1971. Basionym: trib. *Hortonieae* Perkins & Gilg in Engler, Pflanzenr. 4: 12, 14. 21 Jun 1901. Type: *Hortonia* Wight & Arn.

Huaceae A. Chev. in Rev. Int. Bot. Appl. Agric. Trop. 27: 28. Jan-Feb 1947. Type: *Hua* Pierre ex De Wild.

Hugoniaceae Arn. in Wight & Arnott, Prodr. Fl. Ind. Orient.: 71. 10 Oct 1834. Type: *Hugonia* L.

Humbertiaceae Pichon in Notul. Syst. (Paris) 13: 23. Jul-Sep 1947, *nom. cons.* Type: *Humbertia* Lam.

Humiriaceae A. Juss. in Saint-Hilaire, Fl. Bras. Merid. 2: 87. 10

Oct 1829, *nom. cons.* Type: *Humiria* Aubl., *orth. cons.*

Hydnoraceae C. Agardh, Aphor. Bot.: 88. 21 Dec 1821, *nom. cons.* Type: *Hydnora* Thunb.

Hydrangeaceae Dumort., Anal. Fam. Pl.: 36, 38. 1829, *nom. cons.* (to be conserved against *Philadelphaceae*). Type: *Hydrangea* L.

Hydrastidaceae Martinov, Tekhno-Bot. Slovar: 318. 1820. Type: Hydrastis L.

Hydrocotylaceae (Drude) Hyl. [Nomenkl. Stud. Nord. Gefässpfl.] in Uppsala Univ. Årsskr. 1945(7): 20. Jun-Dec 1945, *nom. cons.* *Basionym: subfam. Hydrocotyloideae* Drude in Engler & Prantl, Nat. Pflanzenfam. 3(8): 114, 116. Dec 1897. Type: *Hydrocotyle* L.

Hydropeltidaceae (DC.) Dumort., Comment. Bot.: 64. Nov-Dec 1822. Basionym: trib. *Hydropeltideae* DC., Syst. Nat. 2: 36. Mai 1921. Type: *Hydropeltis* Michx. [= *Brasenia* Schreb.]

Hydrophyllaceae R. Br. in Bot. Reg. 3: sub t. 242. 1 Dec 1817, *nom. cons.* Type: *Hydrophyllum* L.

Hydrostachyaceae Engl., Syllabus, ed. 2: 125. Mai 1898, *nom. cons.* Type: *Hydrostachys* Thouars

Hymenocardiaceae Airy Shaw in Kew Bull. 18: 261. 8 Dec 1965. Type: *Hymenocardia* Wall. ex Lindl.

Hypecoaceae (Dumort.) Willk. & Lange, Prodr. Fl. Hispan. 3: 875. Apr-Mai 1880. Basionym: trib. *Hypecoeae* Dumort., Fl. Belg.: 130. 1827-1830. Type: *Hypecoum* L.

Hypericaceae Juss., Gen. Pl.: 254. 4 Aug 1789, *nom. cons.* Type: Hypericum L.

Hypseocharitaceae Wedd., Chlor. Andina 2: 288. Nov 1861. Type: *Hypseocharis* J. Rémy

Icacinaceae (Benth.) Miers in Ann. Mag. Nat. Hist., ser. 2, 8: 174. Sep 1851, *nom. cons.* Basionym: trib. *Icacineae* Benth. in Trans. Linn. Soc. London 18: 679. 1841. Type: *Icacina* A. Juss.

Idiospermaceae S. T. Blake in Contr. Queensland Herb. 12: 3. 4 Sep 1972. Type: *Idiospermum* S. T. Blake

Illecebraceae R. Br., Prodr.: 413. 27 Mar 1810, *nom. cons.* Type: *Illecebrum* L.

Illiciaceae (DC.) A. C. Sm. in Sargentia 7: 8. 28 Nov 1947, *nom. cons.* Basionym: trib. *Illicieae* DC., Prodr. 1: 77. Jan 1824. Type: *Illicium* L.

Irvingiaceae (Engl.) Exell & Mendonça, Consp. Fl. Angol. 1: 279, 395. 20 Aug 1951, *nom. cons.* Basionym: *subfam. Irvingioideae* Engl., Veg. Erde 9 (3, 1): 765. Feb-Sep 1915. Type: *Irvingia* Hook. f.

Iteaceae J. Agardh, Theoria Syst. Pl.: 151. Apr-Sep 1858, *nom. cons.* Type: *Itea* L.

Ixonanthaceae (Benth.) Exell & Mendonça in Bol. Soc. Brot., ser. 2, 25: 105. 1951, *nom. cons.* Basionym: trib. *Ixonantheae* Benth. in Bentham & Hooker, Gen. Pl. 1:

242, 245. 7 Aug 1862. Type: *Ixonanthes* Jack

Juglandaceae A. Rich. ex Kunth in Ann. Sci. Nat. (Paris) 2: 343. Jun 1824, *nom. cons.* Type: *Juglans* L.

Julianiaceae Hemsl. in J. Bot. 44: 379. Oct 1906, *nom. cons.* Type: *Juliania* Schltdl. non La Llave & Lex. (*Amphipterygium* Standl.)

Justiciaceae Raf., Fl. Tellur. 4: 60. medio 1838. Type: *Justicia* L.

Kiggelariaceae Link, Handbuch 2: 221. Jan-Aug 1831. Type: *Kiggelaria* L.

Kingdoniaceae A. S. Foster ex Airy Shaw in Kew Bull. 18: 262. 8 Dec 1965. Type: *Kingdonia* Balf. f. & W. W. Sm.

Kirkiaceae (Engl.) Takht., Sist. Filog. Cvetk. Rast. [Syst. Phylog. Magnolioph.]: 321. 11 Jan - 1 Mar 1967. Basionym: trib. *Kirkieae* Engl. in Engler & Prantl, Nat. Pflanzenfam. 3(4): 208, 226. Apr 1896. Type: *Kirkia* Oliv.

Koeberliniaceae Engl. in Engler & Prantl, Nat. Pflanzenfam. 3(6): 319. 14 Mai 1895, *nom. cons.* Type: *Koeberlinia* Zucc.

Krameriaceae Dumort., Anal. Fam. Pl.: 20, 23. 1829, *nom. cons.* Type: *Krameria* L.

Labiatae Juss., Gen. Pl.: 110. 4 Aug 1789, *nom. cons.;* nom. alt.: *Lamiaceae.* Type: *Lamium* L.

Lacistemataceae Mart., Nov. Gen. Sp. Pl. 1: 158. Jan-Mar 1826, *nom. cons.* Type: *Lacistema* Sw.

Lactoridaceae Engl. in Engler & Prantl, Nat. Pflanzenfam. 3(2): 19. Feb 1888, *nom. cons.* Type: *Lactoris* Phil.

Lamiaceae Lindl., Nat. Syst. Bot., ed. 2: 275. Jul 1836, *nom. cons.;* nom. alt.: *Labiatae.* Type: *Lamium* L.

Lardizabalaceae Decne. in Arch. Mus. Hist. Nat. 1: 185. 1839, *nom. cons.* Type: *Lardizabala* Ruiz & Pav.

Latraeophileaceae – see *Helosaceae*

Lauraceae Juss., Gen. Pl.: 80. 4 Aug 1789, *nom. cons.* Type: *Laurus* L.

Lecythidaceae Poit. in Mém. Mus. Hist. Nat. 13: 143. 1825, *nom. cons.* Type: *Lecythis* Loefl.

Ledocarpaceae Meyen, Reise 1: 308. 25-31 Mai 1834. Type: *Ledocarpon* Desf. [= *Balbisia* Cav., *nom. cons.*]

Leeaceae (DC.) Dumort., Anal. Fam. Pl.: 21, 27. 1829, *nom. cons.* Basionym: trib. *Leeeae* DC., Prodr. 1: 635. Jan 1824. Type: *Leea* D. Royen ex L., *nom. cons.*

Leguminosae Juss., Gen. Pl.: 345. 4 Aug 1789, *nom. cons.;* nom. alt.: *Fabaceae.* Type: *Faba* Hill [= *Vicia* L.]

Leitneriaceae Benth. in Bentham & Hooker, Gen. Pl. 3: vi, 396. 7 Feb 1880, *nom. cons.* Type: *Leitneria* Chapm.

Lennoaceae Solms in Abh. Naturf. Ges. Halle 11: 174. 7 Jan 1870, *nom. cons.* Type: *Lennoa* La Llave & Lex.

Lentibulariaceae Rich. in Poiteau & Turpin, Fl. Paris. 1, ed. 4°: 26 [ed. f°: 23]. 1808, *nom. cons.* Type: *Lentibularia* Ség. (*Utricularia* L.)

Leonticaceae Bercht. & J. Presl, Přir. Rostlin 1: 217. 1820 (to be rejected in favour of *Podophyllaceae*). Type: *Leontice* L.

Lepidobotryaceae J. Léonard in Bull. Jard. Bot. Etat 20: 38. Jun 1950, *nom. cons.* Type: *Lepidobotrys* Engl.

Leptospermaceae F. Rudolphi, Syst. Orb. Veg.: 55. 1830. Type: *Leptospermum* J. R. Forst. & G. Forst.

Lepuropetalaceae (Engl.) Nakai, Chosakuronbun Mokuroku [Ord. Fam. Trib. Nov.]: 243. 20 Jul 1943. Basionym: subfam. *Lepuropetaloideae* Engl. in Engler & Prantl, Nat. Pflanzenfam., ed. 2, 18a: 108, 178. [seors. impr.] 1928. Type: *Lepuropetalon* Elliott

Limnanthaceae R. Br. in London Edinburgh Philos. Mag. & J. Sci. 3: 70. 1833, *nom. cons.* Type: *Limnanthes* R. Br., *nom. cons.*

Limoniaceae Ser., Fl. Pharm.: 456. 1851, *nom. cons. prop.* Type: *Limonium* Mill., *nom. cons.*

Linaceae DC. ex Gray, Nat. Arr. Brit. Pl. 2: 622, 639. 1 Nov 1821, *nom. cons.* Type: *Linum* L.

Liriodendraceae F. A. Barkley in Phytologia 32: 304. 13 Nov 1975. Type: *Liriodendron* L.

Lissocarpaceae Gilg in Engler, Syllabus, ed. 9 & 10: 324. Nov-Dec 1924, *nom. cons.* Type: *Lissocarpa* Benth.

Loasaceae Dumort., Comment. Bot.: 58. Nov-Dec 1822, *nom. cons.* Type: *Loasa* Adans.

Lobeliaceae R. Br. [Observ. Compositae] in Trans. Linn. Soc. London 12: 133. [seors. impr.] Aug 1817, *nom. cons.* Type: *Lobelia* L.

Loganiaceae R. Br. ex Mart., Nov. Gen. Sp. Pl. 2: 133. Jan-Jun 1827, *nom. cons.* (to be conserved against *Strychnaceae;* to be treated as having priority over *Potaliaceae* and *Spigeliaceae*). Type: *Logania* R. Br., *nom. cons.*

Lophiraceae Loudon, Hort. Brit.: 513. 1-7 Sep 1830. Type: *Lophira* Banks ex C. F. Gaertn.

Lophophytaceae Horan., Char. Ess. Fam.: 22. 1847. Type: *Lophophytum* Schott & Endl.

Lophopyxidaceae (Engl.) H. Pfeiff. in Revista Sudamer. Bot. 10: 4. Apr 1951. Basionym: subfam. *Lophopyxidoideae* Engl. in Engler & Prantl, Nat. Pflanzenfam. 3(5): 257. Mai 1893. Type: *Lophopyxis* Hook. f.

Loranthaceae Juss. in Ann. Mus. Natl. Hist. Nat. 12: 292. 1808, *nom. cons.* (to be conserved against *Viscaceae*). Type: *Loranthus* Jacq., *nom. cons.*

Lythraceae J. St.-Hil., Expos. Fam. Nat. 2: 175. Feb-Apr 1805, *nom. cons.* Type: *Lythrum* L.

Magnoliaceae Juss., Gen. Pl.: 280. 4 Aug 1789, *nom. cons.* Type: *Magnolia* L.

Malaceae Small ex Britton & Small, Fl. S.E. U.S.: 529. 22 Jul 1903, *nom. cons.* Type: *Malus* Mill.

Malesherbiaceae D. Don in Edinburgh New Philos. J. 2: 321. 1827, *nom. cons.* Type: *Malesherbia* Ruiz & Pav.

Malpighiaceae Juss., Gen. Pl.: 252. 4 Aug 1789, *nom. cons.* Type: *Malpighia* L.

Malvaceae Juss., Gen. Pl.: 271. 4 Aug 1789, *nom. cons.* Type: *Malva* L.

Marcgraviaceae Choisy in Candolle, Prodr. 1: 565. Jan 1824, *nom. cons.* Type: *Marcgravia* L.

Martyniaceae Stapf in Engler & Prantl, Nat. Pflanzenfam. 4(3b): 265. 12 Mar 1895, *nom. cons.* Type: *Martynia* L.

Mastixiaceae Calest. in Webbia 1: 94. 10 Mai 1905. Type: *Mastixia* Blume

Medusagynaceae Engl. & Gilg in Engler, Syllabus, ed. 9 & 10: 280. Nov-Dec 1924, *nom. cons.* Type: *Medusagyne* Baker

Medusandraceae Brenan in Kew Bull. 1952: 228. 25 Jul 1952, *nom. cons.* Type: *Medusandra* Brenan

Melanophyllaceae Takht. ex Airy Shaw in Kew Bull. 26: 492. 18 Apr 1972. Type: *Melanophylla* Baker

Melastomataceae Juss., Gen. Pl.: 328. 4 Aug 1789, *nom. cons.* Type: *Melastoma* L.

Meliaceae Juss., Gen. Pl.: 263. 4 Aug 1789, *nom. cons.* Type: *Melia* L.

Melianthaceae Link, Handbuch 2: 322. Jan-Aug 1831, *nom. cons.* Type: *Melianthus* L.

Meliosmaceae Endl., Ench. Bot.: 564. 15-21 Aug 1841 (to be rejected in favour of *Sabiaceae*). Type: *Meliosma* Blume

Memecylaceae DC., Prodr. 3: 5. Mar 1828. Type: *Memecylon* L.

Mendonciaceae Bremek. in Proc. Kon. Ned. Akad. Wetensch., C, 56: 540. 27 Apr 1954. Type: *Mendoncia* Vell. ex Vand.

Menispermaceae Juss., Gen. Pl.: 284. 4 Aug 1789, *nom. cons.* Type: *Menispermum* L.

Menyanthaceae (Dumort.) Dumort., Anal. Fam. Pl.: 20, 25. 1829, *nom. cons.* Basionym: trib. *Menyantheae* Dumort., Fl. Belg.: 52. 1827-1830. Type: *Menyanthes* L.

Mesembryanthemaceae Fenzl in Ann. Wiener Mus. Naturgesch. 1: 349. 1836, *nom. cons.* Type: *Mesembryanthemum* L.

Metteniusaceae Schnizl., Iconogr. Fam. Regn. Veg. 2: sub t. 142*. 1860-1870. Type: *Metteniusa* H. Karst.

Meyeniaceae Sreem. in Phytologia 37: 412. 22 Oct 1977. Type: *Meyenia* Nees

Mimosaceae R. Br. in Flinders, Voy. Terra Austr. 2: 551. 19 Jul 1814, *nom. cons.* Type: *Mimosa* L.

Misodendraceae J. Agardh, Theoria Syst. Pl.: 236. Apr-Sep 1858, *nom. cons.* Type: *Misodendrum* Banks ex DC.

Mitrastemonaceae Makino in Bot. Mag. (Tokyo) 25: 252. Dec 1911, *nom. cons.* Type: *Mitrastemon* Makino

Molluginaceae Hutch., Fam. Fl. Pl. 1: 128. Feb-Mar 1926, *nom. cons.* Type: *Mollugo* L.

Monimiaceae Juss. in Ann. Mus. Natl. Hist. Nat. 14: 133. 1809, *nom. cons.* Type: *Monimia* Thouars

Monotaceae (Gilg) Maury ex Takht., Sist. Magnoliof. [Syst. Magnolioph.]: 129. 31 Mar - 1 Jul 1987. Basionym: subfam. *Monotoideae* Gilg in Engler & Prantl, Nat. Pflanzenfam., ed. 2, 21: 250, 268. Dec 1925. Type: *Monotes* A. DC.

Monotropaceae Nutt., Gen. N. Amer. Pl. 1: 272. 14 Jul 1818, *nom. cons.* (rejected in favour of *Pyrolaceae*). Type: *Monotropa* L.

Montiniaceae Nakai, Chosakuronbun Mokuroku [Ord. Fam. Trib. Nov.]: 243. 20 Jul 1943, *nom. cons.* Type: *Montinia* Thunb.

Moraceae Link, Handbuch 2: 444. Jan-Aug 1831, *nom. cons.* Type: *Morus* L.

Morinaceae Raf. in Ann. Gén. Sci. Phys. 6: 88. 1820. Type: *Morina* L.

Moringaceae R. Br. ex Dumort., Anal. Fam. Pl.: 43, 48. 1829, *nom. cons.* Type: *Moringa* Adans.

Myoporaceae R. Br., Prodr.: 514. 27 Mar 1810, *nom. cons.* Type: *Myoporum* Sol. ex G. Forst.

Myricaceae Blume, Fl. Javae 17-18: 3. 17 Oct 1829, *nom. cons.* Type: *Myrica* L.

Myriophyllaceae Schultz Sch., Nat. Syst. Pflanzenr.: 324. 1832. Type: *Myriophyllum* L.

Myristicaceae R. Br., Prodr.: 399. 27 Mar 1810, *nom. cons.* Type: *Myristica* Gronov., *nom. cons.*

Myrothamnaceae Nied. in Engler & Prantl, Nat. Pflanzenfam. 3(2a): 103. Mar 1891, *nom. cons.* Type: *Myrothamnus* Welw.

Myrsinaceae R. Br., Prodr.: 532. 27 Mar 1810, *nom. cons.* Type: *Myrsine* L.

Myrtaceae Juss., Gen. Pl.: 322. 4 Aug 1789, *nom. cons.* Type: *Myrtus* L.

Mystropetalaceae Hook. f. in Lindley, Veg. Kingd., ed. 3: 94. Aug 1853. Type: *Mystropetalon* Harv.

Nandinaceae Horan., Prim. Lin. Syst. Nat.: 90. Jul-Dec 1834. Type: *Nandina* Thunb.

Napoleonaeaceae A. Rich. in Bory, Dict. Class. Hist. Nat. 11: 432. 10 Feb 1827. Type: *Napoleonaea* P. Beauv.

Naucleaceae (DC.) Wernham in New Phytol. 11: 225. 1911. Basionym: subtrib. *Naucleinae* DC., Prodr. 4: 342, 343. Sep 1830. Type: *Nauclea* L.

Nelsoniaceae (Nees) Sreemadhaven in Phytologia 37: 412. 22 Oct 1977. Basionym: trib. *Nelsonieae*

Nees in Wallich, Pl. Asiat. Rar. 3: 74. 15 Aug 1832. Type: *Nelsonia* R. Br.

Nelumbonaceae (DC.) Dumort., Anal. Fam. Pl.: 53. 1829, *nom. cons.* Basionym: trib. *Nelumboneae* DC., Syst. Nat. 2: 43. Mai 1821. Type: *Nelumbo* Adans.

Nemacladaceae Nutt. in Trans. Amer. Philos. Soc., ser. 2, 8: 254. 15 Dec 1842. Type: *Nemacladus* Nutt.

Nepenthaceae Dumort., Anal. Fam. Pl.: 14, 16. 1829, *nom. cons.* Type: *Nepenthes* L.

Nesogenaceae Marais in Kew Bull. 35: 798. 28 Apr 1981. Type: *Nesogenes* A. DC.

Neuradaceae Link, Handbuch 2: 97. Jan-Aug 1831, *nom. cons. prop.* Type: *Neurada* L.

Nitrariaceae Bercht. & J. Presl, Přir. Rostlin 1: 238. 1820. Type: *Nitraria* L.

Nolanaceae Dumort., Anal. Fam. Pl.: 20, 24. 1829, *nom. cons.* Type: *Nolana* L. ex L. f.

Nothofagaceae Kuprian. in Sladkov, First Int. Conf. Palinol. Reports Soviet Palinol.: 21. 1962. Type: *Nothofagus* Blume, *nom. cons.*

Nyctaginaceae Juss., Gen. Pl.: 90. 4 Aug 1789, *nom. cons.* Type: *Nyctago* Juss., nom. illeg. (*Mirabilis* L.)

Nymphaeaceae Salisb. in Ann. Bot. (Koenig & Sims) 2: 70. Jun 1805, *nom. cons.* Type: *Nymphaea* L., *nom. cons.*

Nyssaceae Juss. ex Dumort., Anal. Fam. Pl.: 13. 1829, *nom. cons.* (*Cornaceae* to be treated as having priority). Type: *Nyssa* L.

Ochnaceae DC. in Ann. Mus. Natl. Hist. Nat. 17: 410. Jul-Aug 1811, *nom. cons.* Type: *Ochna* L.

Octoknemaceae Engl. in Bot. Jahrb. Syst. 43: 177. 23 Feb 1909, *nom. cons.* Type: *Octoknema* Pierre

Oenotheraceae – see *Onagraceae*

Oftiaceae Takht. & Reveal in Phytologia 74: 284. Apr 1993. Type: *Oftia* Adans.

Olacaceae Mirb. ex DC., Prodr. 1: 531. Jan 1824, *nom. cons.* Type: *Olax* L.

Oleaceae Hoffmanns. & Link, Fl. Portug. 1: 385. 1813-1820, *nom. cons.* Type: *Olea* L.

Oliniaceae Harv. & Sond., Fl. Cap. 2: ix. 15-31 Oct 1862, *nom. cons.* Type: *Olinia* Thunb., *nom. cons.*

Onagraceae Juss., Gen. Pl.: 317. 4 Aug 1789, *nom. cons.* Type: *Onagra* Hill (*Oenothera* L.)

Oncothecaceae Kobuski ex Airy Shaw in Kew Bull. 18: 264. 8 Dec 1965. Type: *Oncotheca* Baill.

Opiliaceae (Benth.) Valeton, Crit. Overz. Olacin.: 136. 7 Jul 1886, *nom. cons.* Basionym: trib. *Opilieae* Benth. in Trans. Linn. Soc. London 18: 679. 1841. Type: *Opilia* Roxb.

Orobanchaceae Vent., Tabl. Règne Vég. 2: 292. 5 Mai 1799, *nom. cons.* Type: *Orobanche* L.

Oxalidaceae R. Br. in Tuckey, Narr. Exped. Zaire: 433. 5 Mar 1818, *nom. cons.* Type: *Oxalis* L.

Oxystylidaceae Hutch., Evol. Phylog. Fl. Pl.: 516. Jan-Sep 1969. Type: *Oxystylis* Torr. & Frém.

Paeoniaceae F. Rudolphi, Syst. Orb. Veg.: 61. 1830, *nom. cons.* Type: *Paeonia* L.

Paivaeusaceae – see *Picrodendraceae*

Pandaceae Engl. & Gilg in Engler, Syllabus, ed. 7: 223. Oct 1912 - Mar 1913, *nom. cons.* Type: *Panda* Pierre

Papaveraceae Juss., Gen. Pl.: 235. 4 Aug 1789, *nom. cons.* Type: *Papaver* L.

Papilionaceae Giseke, Prael. Ord. Nat. Pl.: 415. Apr 1792, *nom. cons.;* nom. alt.: *Fabaceae.* Type: *Faba* Hill [= *Vicia* L.]

Paracryphiaceae Airy Shaw in Kew Bull. 18: 265. 8 Dec 1965. Type: *Paracryphia* Baker f.

Parnassiaceae Gray, Nat. Arr. Brit. Pl. 2: 623, 670. 1 Nov 1821, *nom. cons.* Type: *Parnassia* L.

Passifloraceae Juss. ex Kunth in Humboldt & al., Nov. Gen. Sp. 2, ed. 4°: 126 [ed. f°: 100]. 8 Dec 1817, *nom. cons.* Type: *Passiflora* L.

Pedaliaceae R. Br., Prodr.: 519. 27 Mar 1810, *nom. cons.* Type: *Pedalium* D. Royen ex L.

Peganaceae (Engl.) Tiegh. ex Takht., Sist. Magnoliof. [Syst. Magnolioph.]: 178. 31 Mar - 1 Jul 1987. Basionym: subfam. *Peganoideae* Engl. in Engler & Prantl, Nat. Pflanzenfam. 3(4): 90. Dec 1890. Type: *Peganum* L.

Pellicieraceae (Triana & Planch.) L. Beauvis. ex Bullock in Taxon 8: 182. 24 Jul 1959. Basionym: trib. *Pelliciereae* Triana & Planch. in Ann. Sci. Nat., Bot., ser. 4, 17: 380. Jun 1862. Type: *Pelliciera* Planch. & Triana ex Benth. & Hook. f.

Penaeaceae Sweet ex Guill. in Bory, Dict. Class. Hist. Nat. 13: 171. 1 Mar 1828, *nom. cons.* Type: *Penaea* L.

Pentadiplandraceae Hutch. & Dalziel, Fl. W. Trop. Afr. 1: 461. Jul 1928. Type: *Pentadiplandra* Baill.

Pentaphragmataceae J. Agardh, Theoria Syst. Pl.: 95. Apr-Sep 1858, *nom. cons.* Type: *Pentaphragma* Wall. ex G. Don

Pentaphylacaceae Engl. in Engler & Prantl, Nat. Pflanzenfam., Nachtr. 1: 214. Oct 1897, *nom. cons.* Type: *Pentaphylax* Gardner & Champ.

Pentastemonaceae Duyfjes in Blumea 36: 552. 9 Jun 1992. Type: *Pentastemona* Steenis

Penthoraceae Rydb. ex Britton, Man. Fl. N. States: 475. Oct 1901, *nom. cons.* Type: *Penthorum* L.

Peperomiaceae A. C. Sm., Fl. Vit. Nova 2: 75. 26 Oct 1981. Type: *Peperomia* Ruiz & Pav.

Peraceae Klotzsch in Monatsber. Königl. Preuss. Akad. Wiss. Berlin 1859: 241, 246. 1859. Type: *Pera* Mutis

Peridiscaceae Kuhlm. in Arq. Serv. Florest. 3: 4. 1950, *nom. cons.* Type: *Peridiscus* Benth.

Periplocaceae Schltr. in Schumann & Lauterbach, Nachtr. Fl. Schutzgeb. Südsee: 351. Nov 1905, *nom. cons.* Type: *Periploca* L.

Peripterygiaceae – see *Cardiopteridaceae*

Petiveriaceae C. Agardh, Aphor. Bot.: 221. 13 Jun 1824. Type: *Petiveria* L.

Phellinaceae (Loes.) Takht., Sist. Filog. Cvetk. Rast. [Syst. Phylog. Magnolioph.]: 374. 11 Jan - 1 Mar 1967. Basionym: trib. *Phellineae* Loes. in Nova Acta Acad. Caes. Leop.-Carol. German. Nat. Cur. 78: 504. 1 Mar 1901. Type: *Phelline* Labill.

Philadelphaceae Martinov, Tekhno-Bot. Slovar: 478. 1820 (to be rejected in favour of *Hydrangeaceae*). Type: *Philadelphus* L.

Phrymaceae Schauer in Candolle, Prodr. 11: 520. 25 Nov 1847, *nom. cons.* Type: *Phryma* L.

Phyllanthaceae J. Agardh, Theoria Syst. Pl.: 249. Apr-Sep 1858. Type: *Phyllanthus* L.

Phyllonomaceae Small in N. Amer. Fl. 22(1-6): 2. 22 Mai 1905. Type: *Phyllonoma* Willd. ex Schult.

Physenaceae Takht. in Bot. Žurn. [Moscow & Leningrad] 70: 1692. 13 Dec 1985. Type: *Physena* Noronha ex Thouars

Phytolaccaceae R. Br. in Tuckey, Narr. Exped. Zaire: 454. 5 Mar 1818, *nom. cons.* Type: *Phytolacca* L.

Picrodendraceae Small ex Britton & Millsp., Bahama Fl.: 102. 26 Jun 1920, *nom. cons.* Type: *Picrodendron* Griseb., *nom. cons.*

Piperaceae C. Agardh, Aphor. Bot.: 201. 13 Jun 1824, *nom. cons.* Type: *Piper* L.

Pistaciaceae Mart. ex Perleb, Clav. Class.: 31. Jan-Mar 1838. Type: *Pistacia* L.

Pittosporaceae R. Br. in Flinders, Voy. Terra Austr. 2: 542. 19 Jul 1814, *nom. cons.* Type: *Pittosporum* Banks ex Sol., *nom. cons.*

Plagiopteraceae Airy Shaw in Kew Bull. 18: 266. 8 Dec 1965. Type: *Plagiopteron* Griff.

Plantaginaceae Juss., Gen. Pl.: 89. 4 Aug 1789, *nom. cons.* Type: *Plantago* L.

Platanaceae Lestiboudois ex Dumort., Anal. Fam. Pl.: 11, 12. 1829, *nom. cons.* Type: *Platanus* L.

Platystemonaceae (W. R. Ernst) A. C. Sm. in J. Indian Bot. Soc. 50A: 222. Dec 1971. Basionym: subfam. *Platystemonoideae* W. R. Ernst in J. Arnold Arbor. 43: 317. 10 Jul 1962. Type: *Platystemon* Benth.

Plocospermataceae Hutch., Fam. Fl. Pl., ed. 3: 469. Jan-Jul 1973. Type: *Plocosperma* Benth.

Plumbaginaceae Juss., Gen. Pl.: 92. 4 Aug 1789, *nom. cons.* Type: *Plumbago* L.

Podoaceae Baill. ex Franch., Pl. Delavay.: 145. Mai 1889. Type: *Podoon* Baill. [= *Dobinea* Buch.-Ham. ex D. Don]

Podophyllaceae DC., Syst. Nat. 2: 31. Mai 1821, *nom. cons.* (to be conserved against *Leonticaceae*). Type: *Podophyllum* L.

Podostemaceae Rich. ex C. Agardh, Aphor. Bot.: 125. 19 Jun 1822, *nom. cons.* Type: *Podostemum* Michx.

Polemoniaceae Juss., Gen. Pl.: 136. 4 Aug 1789, *nom. cons.* Type: *Polemonium* L.

Polygalaceae R. Br. in Flinders, Voy. Terra Austr. 2: 542. 19 Jul 1814, *nom. cons.* Type: *Polygala* L.

Polygonaceae Juss., Gen. Pl.: 82. 4 Aug 1789, *nom. cons.* Type: *Polygonum* L.

Polyosmaceae Blume, Mus. Bot. 1: 258. init. 1851. Type: *Polyosma* Blume

Portulacaceae Juss., Gen. Pl.: 312. 4 Aug 1789, *nom. cons.* Type: *Portulaca* L.

Potaliaceae Mart., Nov. Gen. Sp. Pl. 2: 133. Jan-Jun 1827 (*Loganiaceae* to be treated as having priority). Type: *Potalia* Aubl.

Pottingeriaceae (Engl.) Takht., Sist. Magnoliof. [Syst. Magnolioph.]: 212. 31 Mar - 1 Jul 1987. Basionym: trib. *Pottingerieae* Engl. in Engler & Prantl, Nat. Pflanzenfam., ed. 2, 18a: 111, 217. [seors. impr.] 1928. Type: *Pottingeria* Prain

Primulaceae Vent., Tabl. Règne Vég. 2: 285. 5 Mai 1799, *nom. cons.* Type: *Primula* L.

Prionotaceae Hutch., Evol. Phylog. Fl. Pl.: 306. Jan-Sep 1969. Type: *Prionotes* R. Br.

Proteaceae Juss., Gen. Pl.: 78. 4 Aug 1789, *nom. cons.* Type: *Protea* L., *nom. cons.*

Psiloxylaceae Croizat, Principia Bot.: 604, 1161, 1164. Jun 1961. Type: *Psiloxylon* Thouars ex Tul.

Ptaeroxylaceae J.-F. Leroy in J. Agric. Trop. Bot. Appl. 7: 456. Sep-Oct 1960. Type: *Ptaeroxylon* Eckl. & Zeyh.

Pteridophyllaceae (Murb.) Nakai ex Reveal & Hoogland in Bull. Mus. Natl. Hist. Nat., B, Adansonia 13: 91. 4 Oct 1991. Basionym: subfam. *Pteridophylloideae* Murb. in Kongl. Svenska Vetenskapsakad. Handl., ser. 2, 50(1): 114. 17 Dec 1912. Type: *Pteridophyllum* Siebold & Zucc.

Pterostemonaceae Small in N. Amer. Fl. 22(1-6): 183. 18 Dec 1905, *nom. cons.* Type: *Pterostemon* Schauer

Punicaceae Horan., Prim. Lin. Syst. Nat.: 81. Jul-Dec 1834, *nom. cons.* Type: *Punica* L.

Putranjivaceae Endl., Ench. Bot.: 174. 15-21 Aug 1841. Type: *Putranjiva* Wall.

Pyrolaceae Dumort., Anal. Fam. Pl.: 43, 47, 80. 1829, *nom. cons.* (conserved against *Monotropaceae*). Type: *Pyrola* L.

Quiinaceae Choisy ex Engl. in Martius, Fl. Bras. 12(1): 475-476.

1 Apr 1888, *nom. cons.* Type: *Quiina* Aubl.

Rafflesiaceae Dumort., Anal. Fam. Pl.: 13, 14. 1829, *nom. cons.* (to be conserved against *Cytinaceae*). Type: *Rafflesia* R. Br.

Ranunculaceae Juss., Gen. Pl.: 231. 4 Aug 1789, *nom. cons.* Type: *Ranunculus* L.

Resedaceae DC. ex Gray, Nat. Arr. Brit. Pl. 2: 622, 665. 1 Nov 1821, *nom. cons.* Type: *Reseda* L.

Retziaceae Bartl., Ord. Nat. Pl.: 192. Sep 1830 (to be rejected in favour of *Stilbaceae*). Type: *Retzia* Thunb.

Rhabdodendraceae (Huber) Prance in Bull. Jard. Bot. Etat 38: 141. 30 Jun 1968, *nom. cons.* Basionym: trib. *Rhabdodendreae* Huber in Bol. Mus. Goeldi Paraense Hist. Nat. Ethnogr. 5: 425. 1909. Type: *Rhabdodendron* Gilg & Pilg.

Rhamnaceae Juss., Gen. Pl.: 376. 4 Aug 1789, *nom. cons.* Type: *Rhamnus* L.

Rhizophoraceae R. Br. in Flinders, Voy. Terra Austr. 2: 549. 19 Jul 1814, *nom. cons.* Type: *Rhizophora* L.

Rhodoleiaceae Nakai, Chosakuronbun Mokuroku [Ord. Fam. Trib. Nov.]: 246. 20 Jul 1943. Type: *Rhodoleia* Champ. ex Hook.

Rhoipteleaceae Hand.-Mazz. in Repert. Spec. Nov. Regni Veg. 30: 75. 15 Feb 1932, *nom. cons.* Type: *Rhoiptelea* Diels & Hand.-Mazz.

Rhopalocarpaceae – see *Sphaerosepalaceae*

Rhynchocalycaceae L. A. S. Johnson & B. Briggs in Ann. Missouri Bot. Gard. 71: 732. 19 Apr 1985. Type: *Rhynchocalyx* Oliv.

Rhynchothecaceae Endl., Ench. Bot.: 621. 15-21 Aug 1841. Type: *Rhynchotheca* Ruiz & Pav.

Ribesaceae – see *Grossulariaceae*

Roridulaceae Engl. & Gilg in Engler, Syllabus, ed. 9 & 10: 226. Nov-Dec 1924, *nom. cons.* Type: *Roridula* Burm. f. ex L.

Rosaceae Juss., Gen. Pl.: 334. 4 Aug 1789, *nom. cons.* Type: *Rosa* L.

Rousseaceae DC., Prodr. 7(2): 521. Dec 1839. Type: *Roussea* Sm.

Rubiaceae Juss., Gen. Pl.: 196. 4 Aug 1789, *nom. cons.* Type: *Rubia* L.

Rutaceae Juss., Gen. Pl.: 296. 4 Aug 1789, *nom. cons.* Type: *Ruta* L.

Sabiaceae Blume, Mus. Bot. 1: 368. 1851, *nom. cons.* (to be conserved against *Meliosmaceae*). Type: *Sabia* Colebr.

Saccifoliaceae Maguire & Pires in Mem. New York Bot. Gard. 29: 242. 14 Jun 1978. Type: *Saccifolium* Maguire & Pires

Salazariaceae F. A. Barkley in Phytologia 32: 304. 13 Nov 1975. Type: *Salazaria* Torr.

Salicaceae Mirb., Elém. Physiol. Vég. Bot. 2: 905. 1815, *nom. cons.* Type: *Salix* L.

Salicorniaceae J. Agardh, Theoria Syst. Pl.: 357. Apr-Sep 1858. Type: *Salicornia* L.

Salpiglossidaceae (Benth.) Hutch., Evol. Phylog. Fl. Pl.: 631. Jan-Sep 1969. Basionym: trib. *Salpiglossideae* Benth. in Edwards's Bot. Reg. 21: sub t. 1770. 1 Jun 1835. Type: *Salpiglossis* Ruiz & Pav.

Salsolaceae Moq. in Candolle, Prodr. 13(2): 41. 5 Mai 1849. Type: *Salsola* L.

Salvadoraceae Lindl., Nat. Syst. Bot., ed. 2: 269. Jul 1836, *nom. cons.* Type: *Salvadora* L.

Sambucaceae Batsch ex Borkh., Bot. Wörterbuch 2: 322. 1797 (to be rejected in favour of *Adoxaceae*). Type: *Sambucus* L.

Samydaceae Vent. in Mém. Cl. Sci. Math. Inst. Natl. France 1807(2): 149. 1808, *nom. cons.* (rejected in favour of *Flacourtiaceae*). Type: *Samyda* Jacq., *nom. cons.*

Saniculaceae (Drude) Á. Löve & D. Löve in Preslia 46: 133. 29 Apr 1974. Basionym: subfam. *Saniculoideae* Drude in Engler & Prantl, Nat. Pflanzenfam. 3(8): 115. Dec 1897. Type: *Sanicula* L.

Santalaceae R. Br., Prodr.: 350. 27 Mar 1810, *nom. cons.* Type: *Santalum* L.

Sapindaceae Juss., Gen. Pl.: 246. 4 Aug 1789, *nom. cons.* (to be treated as having priority over *Aceraceae*). Type: *Sapindus* L.

Sapotaceae Juss., Gen. Pl.: 151. 4 Aug 1789, *nom. cons.* Type: *Sapota* Mill. (*Achras* L., *nom. rejic.*) [= *Manilkara* Adans., *nom. cons.*]

Sarcolaenaceae Caruel in Atti Reale Accad. Lincei Mem. Cl. Sci. Fis., ser. 3, 10: 226, 248. 1881, *nom. cons.* Type: *Sarcolaena* Thouars

Sarcophytaceae A. Kern., Pflanzenleben 2: 708. 1891. Type: *Sarcophyte* Sparrm.

Sarcospermataceae H. J. Lam in Bull. Jard. Bot. Buitenzorg, ser. 3, 7: 248. Feb 1925, *nom. cons.* Type: *Sarcosperma* Hook. f.

Sargentodoxaceae Stapf ex Hutch., Fam. Fl. Pl. 1: 100. Feb-Mar 1926, *nom. cons.* Type: *Sargentodoxa* Rehder & E. H. Wilson

Sarraceniaceae Dumort., Anal. Fam. Pl.: 53. 1829, *nom. cons.* Type: *Sarracenia* L.

Saurauiaceae J. Agardh, Theoria Syst. Pl.: 110. Apr-Sep 1858, *nom. cons.* (rejected in favour of *Actinidiaceae*). Type: *Saurauia* Willd., *orth. cons.*

Saururaceae Rich. ex E. Mey., Houttuynia: 20. 5 Dec 1827, *nom. cons.* Type: *Saururus* L.

Sauvagesiaceae (Ging. ex DC.) Dumort., Anal. Fam. Pl.: 44, 49. 1829. Basionym: trib. *Sauvagesieae* Ging. ex DC., Prodr. 1: 315. Jan 1824. Type: *Sauvagesia* L.

Saxifragaceae Juss., Gen. Pl.: 308. 4 Aug 1789, *nom. cons.* Type: *Saxifraga* L.

Schisandraceae Blume, Fl. Javae 32-33: 3. 25 Jun 1830, *nom. cons.* Type: *Schisandra* Michx., *nom. cons.*

Schoepfiaceae Blume, Mus. Bot. 1: 175. Apr 1850. Type: *Schoepfia* Schreb.

Sclerophylacaceae Miers in London J. Bot. 7: 57. Jan-Feb 1848. Type: *Sclerophylax* Miers

Scrophulariaceae Juss., Gen. Pl.: 117. 4 Aug 1789, *nom. cons.* Type: *Scrophularia* L.

Scyphostegiaceae Hutch., Fam. Fl. Pl. 1: 229. Feb-Mar 1926, *nom. cons.* Type: *Scyphostegia* Stapf

Scytopetalaceae Engl. in Engler & Prantl, Nat. Pflanzenfam., Nachtr. 1: 242. Oct 1897, *nom. cons.* Type: *Scytopetalum* Pierre ex Engl.

Selaginaceae Choisy [Mém. Sélag.: 19] in Mém. Soc. Phys. Genève 2: 89. 1823, *nom. cons.* Type: *Selago* L.

Simaroubaceae DC. in Ann. Mus. Natl. Hist. Nat. 17: 422. Jul-Aug 1811, *nom. cons.* Type: *Simarouba* Aubl., *nom. cons.*

Simmondsiaceae (Müll. Arg.) Tiegh. ex Reveal & Hoogland in Bull. Mus. Natl. Hist. Nat., B, Adansonia 12: 206. 24 Nov 1990. Basionym: trib. *Simmondsieae* Müll. Arg. in Candolle, Prodr. 16(1): 9, 22. Nov 1869. Type: *Simmondsia* Nutt.

Siparunaceae (A. DC.) Schodde in Taxon 19: 325. 29 Jun 1970. Basionym: trib. *Siparuneae* A. DC. in Candolle, Prodr. 16(2): 642. Jul 1868. Type: *Siparuna* Aubl.

Siphonodontaceae (Croizat) Gagnep. & Tardieu ex Tardieu in Notul. Syst. (Paris) 14: 102. Jul-Sep 1951, *nom. cons.* Basionym: subfam. *Siphonodontoideae* Croizat in Lilloa 13: 41. 29 Dec 1947. Type: *Siphonodon* Griff.

Sladeniaceae (Gilg & Werderm.) Airy Shaw in Kew Bull. 18: 267. 8 Dec 1965. Basionym: subfam. *Sladenioideae* Gilg & Werderm. in Engler & Prantl, Nat. Pflanzenfam., ed. 2, 21: 41. Dec 1925. Type: *Sladenia* Kurz

Solanaceae Juss., Gen. Pl.: 124. 4 Aug 1789, *nom. cons.* Type: *Solanum* L.

Sonneratiaceae Engl. & Gilg in Engler, Syllabus, ed. 9 & 10: 299. Nov-Dec 1924, *nom. cons.* Type: *Sonneratia* L. f., *nom. cons.*

Sphaerosepalaceae (Warb.) Tiegh. ex Bullock in Taxon 8: 200. 24 Jul 1959. Basionym: trib. *Sphaerosepaleae* Warb. in Engler & Prantl, Nat. Pflanzenfam. 3(6): 310. 14 Mai 1895. Type: *Sphaerosepalum* Baker [= *Rhopalocarpus* Bojer]

Sphenocleaceae Mart. ex DC., Prodr. 7(2): 548. Dec 1839, *nom. cons.* Type: *Sphenoclea* Gaertn., *nom. cons.*

Sphenostemonaceae P. Royen & Airy Shaw in Kew Bull. 27: 325. 23 Oct 1972. Type: *Sphenostemon* Baill.

Spielmanniaceae, nom. illeg. – see *Oftiaceae*

Spigeliaceae Mart., Nov. Gen. Sp. Pl. 2: 132. Jan-Jun 1827 (*Loganiaceae* to be treated as having priority). Type: *Spigelia* L.

Stachyuraceae J. Agardh, Theoria Syst. Pl.: 152. Apr-Sep 1858, *nom. cons.* Type: *Stachyurus* Siebold & Zucc.

Stackhousiaceae R. Br. in Flinders, Voy. Terra Austr. 2: 555. 19 Jul 1814, *nom. cons.* Type: *Stackhousia* Sm.

Staphyleaceae (DC.) Lindl., Syn. Brit. Fl.: 75. Feb 1829, *nom. cons.* Basionym: trib. *Staphyleeae* DC., Prodr. 2: 2. Nov 1825. Type: *Staphylea* L.

Staticaceae – see *Limoniaceae*

Stegnospermataceae (A. Rich.) Nakai in J. Jap. Bot. 18: 108. Mar 1942. Basionym: trib. *Stegnospermateae* A. Rich. in Sagra, Hist. Fis. Cuba, Bot. 10: 308. 1845. Type: *Stegnosperma* Benth.

Sterculiaceae (DC.) Bartl., Ord. Nat. Pl.: 225, 340. Sep 1830, *nom. cons.* (conserved against *Byttneriaceae*). Basionym: trib. *Sterculieae* DC., Prodr. 1: 481. Jan 1824. Type: *Sterculia* L.

Stilaginaceae C. Agardh, Aphor. Bot.: 199. 13 Jun 1824. Type: *Stilago* L. [= *Antidesma* L.]

Stilbaceae Kunth, Handb. Bot.: 393. 1831, *nom. cons.* (to be conserved against *Retziaceae*). Type: *Stilbe* P. J. Bergius

Strasburgeriaceae Engl. & Gilg in Engler, Syllabus, ed. 9 & 10: 282. Nov-Dec 1924, *nom. cons.* Type: *Strasburgeria* Baill.

Strychnaceae DC. ex Perleb, Lehrb. Naturgesch. Pflanzenr.: 220. Sep-Oct 1826 (to be rejected in favour of *Loganiaceae*). Type: *Strychnos* L.

Stylidiaceae R. Br., Prodr.: 565. 27 Mar 1810, *nom. cons.* Type: *Stylidium* Sw. ex Willd., *nom. cons.*

Stylobasiaceae J. Agardh, Theoria Syst. Pl.: 169. Apr-Sep 1858. Type: *Stylobasium* Desf.

Styloceratac eae (Pax) Baill. ex Reveal & Hoogland in Bull. Mus. Natl. Hist. Nat., B, Adansonia 12: 206. 24 Nov 1990. Basionym: trib. *Stylocerateae* Pax in Engler & Prantl, Nat. Pflanzenfam. 3(5): 132, 134. Jun 1892. Type: *Styloceras* Kunth ex A. Juss.

Styracaceae Dumort., Anal. Fam. Pl.: 28, 29. 1829, *nom. cons.* (conserved against *Symplocaceae*). Type: *Styrax* L.

Surianaceae Arn. in Wight & Arnott, Prodr. Fl. Ind. Orient.: 360. 10 Oct 1834, *nom. cons.* Type: *Suriana* L.

Swartziaceae (DC.) Bartl., Ord. Nat. Pl.: 231, 413. Sep 1830. Basionym: trib. *Swartzieae* DC., Prodr. 2: 422. Nov 1825. Type: *Swartzia* Schreb., *nom. cons.*

Symphoremataceae (Meisn.) Moldenke ex Reveal & Hoogland in Bull. Mus. Natl. Hist. Nat., B, Adansonia 13: 92. 4 Oct 1991. Basionym: trib. *Symphoremateae* Meisn., Pl. Vasc. Gen., Tab. Diagn.: 292; Comm.: 200. 25-31 Oct 1840. Type: *Symphorema* Roxb.

Symplocaceae Desf. in Mém. Mus. Hist. Nat. 6: 9. 1820, *nom. cons.* (rejected in favour of *Styracaceae*). Type: *Symplocos* Jacq.

Takhtajaniaceae (J.-F. Leroy) J.-F. Leroy in Adansonia, ser. 2, 20: 20. 30 Mai 1980. Basionym: subfam. *Takhtajanioideae* J.-F. Leroy in Adansonia, ser. 2, 17: 393. 16 Jun 1978. Type: *Takhtajania* Baranova & J.-F. Leroy

Tamaricaceae Link, Enum. Hort. Berol. Alt. 1: 291. 16 Mar - 30 Jun 1821, *nom. cons.* Type: *Tamarix* L.

Tapisciaceae (Pax) Takht., Sist. Magnoliof. [Syst. Magnolioph.]: 171. 31 Mar - 1 Jul 1987. Basionym: subfam. *Tapiscioideae* Pax in Engler & Prantl, Nat. Pflanzenfam. 3(5): 259, 262. Mai 1893. Type: *Tapiscia* Oliv.

Tepuianthaceae Maguire & Steyerm. in Mem. New York Bot. Gard. 32: 8. 20 Mai 1981. Type: *Tepuianthus* Maguire & Steyerm.

Ternstroemiaceae – see *Theaceae*

Tetracarpaeaceae Nakai, Chosakuronbun Mokuroku [Ord. Fam. Trib. Nov.]: 244. 20 Jul 1943. Type: *Tetracarpaea* Hook.

Tetracentraceae A. C. Sm. in J. Arnold Arbor. 26: 135. 16 Apr 1945, *nom. cons.* Type: *Tetracentron* Oliv.

Tetrachondraceae Wettst., Handb. Syst. Bot., ed. 3: 807. Jun-Sep 1924. Type: *Tetrachondra* Petrie ex Oliv.

Tetradiclidaceae (Engl.) Takht., Florist. Reg. World: 333. 4 Sep 1986. Basionym: subfam. *Tetradiclidoideae* Engl. in Engler & Prantl, Nat. Pflanzenfam. 3(4): 355. Jul 1896. Type: *Tetradiclis* Steven ex M. Bieb.

Tetragoniaceae Nakai in J. Jap. Bot. 18: 103. Mar 1942, *nom. cons.* Type: *Tetragonia* L.

Tetramelaceae (Warb.) Airy Shaw in Kew Bull. 18: 267. 8 Dec 1965. Basionym: trib. *Tetrameleae* Warb. in Engler & Prantl, Nat. Pflanzenfam. 3(6a): 152. 18 Apr 1894. Type: *Tetrameles* R. Br.

Tetrameristaceae Hutch., Fam. Fl. Pl., ed. 2: 277. Jan-Jul 1959. Type: *Tetramerista* Miq.

Thalictraceae Raf., Anal. Nat.: 176. Apr-Jul 1815. Type: *Thalictrum* L.

Theaceae D. Don, Prodr. Fl. Nepal.: 224. 26 Jan - 1 Feb 1825, *nom. cons.* Type: *Thea* L. [= *Camellia* L.]

Theligonaceae Dumort., Anal. Fam. Pl.: 15, 17. 1829, *nom. cons.* Type: *Theligonum* L.

Theophrastaceae Link, Handbuch 1: 440. Jan-Aug 1829, *nom. cons.* Type: *Theophrasta* L.

Thomandersiaceae Sreem. in Phytologia 37: 412. 22 Oct 1977. Type: *Thomandersia* Baill.

Thunbergiaceae Bremek. in Proc. Kon. Ned. Akad. Wetensch., C, 56: 540. 27 Apr 1954. Type: *Thunbergia* Retz.

Thymelaeaceae Juss., Gen. Pl.: 76. 4 Aug 1789, *nom. cons.* Type: *Thymelaea* Mill., *nom. cons.*

Ticodendraceae Gómez-Laur. & L. D. Gómez in Ann. Missouri Bot. Gard. 78: 87. 26 Feb 1991. Type:

Ticodendron Gómez-Laur. & L. D. Gómez

Tiliaceae Juss., Gen. Pl.: 289. 4 Aug 1789, *nom. cons.* Type: *Tilia* L.

Torricelliaceae (Wangerin) Hu in Bull. Fan Mem. Inst. Biol., Bot. 5: 311. 1934. Basionym: trib. *Torricellieae* Wangerin in Engler, Pflanzenr. 41: 18, 31. 12 Apr 1910. Type: *Torricellia* DC.

Tovariaceae Pax in Engler & Prantl, Nat. Pflanzenfam. 3(2): 207. Mar 1891, *nom. cons.* Type: *Tovaria* Ruiz & Pav., *nom. cons.*

Trapaceae Dumort., Anal. Fam. Pl.: 36, 39. 1829, *nom. cons.* Type: *Trapa* L.

Trapellaceae Honda & Sakis., Taikou Nippon Shokubutsu Bunraigaku [Syst. Pl. Japon.]: 378. 1930. Type: *Trapella* Oliv.

Tremandraceae R. Br. ex DC., Prodr. 1: 343. Jan 1824, *nom. cons.* Type: *Tremandra* R. Br. ex DC.

Tribelaceae (Engl.) Airy Shaw in Kew Bull. 18: 269. 8 Dec 1965. Basionym: trib. *Tribeleae* Engl. in Engler & Prantl, Nat. Pflanzenfam., ed. 2, 18a: 111, 213. [seors. impr.] 1928. Type: *Tribeles* Phil.

Tribulaceae Trautv., Estestv. Istorija Gub. Kievsk. Učebn. Okr.: 28. 1853. Type: *Tribulus* L.

Trigoniaceae Endl., Ench. Bot.: 570. 15-21 Aug 1841, *nom. cons.* Type: *Trigonia* Aubl.

Trimeniaceae (Perkins & Gilg) Gibbs, Fl. Arfak Mts.: 135. Jul 1917, *nom. cons.* Basionym: trib. *Trimenieae* Perkins & Gilg in Engler, Pflanzenr. 4: 12, 21. 21 Jun 1901. Type: *Trimenia* Seem.

Triplostegiaceae (Höck) Bobrov ex Airy Shaw in Kew Bull. 18: 269. 8 Dec 1965. Basionym: trib. *Triplostegieae* Höck in Bot. Jahrb. Syst. 31: 408. 10 Dec 1901. Type: *Triplostegia* Wall. ex DC.

Tristichaceae Willis in J. Linn. Soc., Bot. 43: 51. 15 Mai 1915. Type: *Tristicha* Thouars

Trochodendraceae Prantl in Engler & Prantl, Nat. Pflanzenfam. 3(2): 21. Feb 1888, *nom. cons.* Type: *Trochodendron* Siebold & Zucc.

Tropaeolaceae Juss. ex DC., Prodr. 1: 683. Jan 1824, *nom. cons.* Type: *Tropaeolum* L.

Turneraceae Kunth ex DC., Prodr. 3: 345. Mar 1828, *nom. cons.* Type: *Turnera* L.

Uapacaceae (Müll. Arg.) Airy Shaw in Kew Bull. 18: 270. 8 Dec 1965. Basionym: subtrib. *Uapacinae* Müll. Arg. in Linnaea 34: 64. Mar 1865. Type: *Uapaca* Baill.

Ulmaceae Mirb., Elém. Physiol. Vég. Bot. 2: 905. 1815, *nom. cons.* Type: *Ulmus* L.

Umbelliferae Juss., Gen. Pl.: 218. 4 Aug 1789, *nom. cons.;* nom. alt.: *Apiaceae.* Type: *Apium* L.

Urticaceae Juss., Gen. Pl.: 400. 4 Aug 1789, *nom. cons.* Type: *Urtica* L.

Vacciniaceae DC. ex Gray, Nat. Arr. Brit. Pl. 2: 394, 404. 1 Nov 1821, *nom. cons.* Type: *Vaccinium* L.

Vahliaceae Dandy in Hutchinson, Fam. Fl. Pl., ed. 2: 461. Jan-Jul 1959. Type: *Vahlia* Thunb., *nom. cons.*

Valerianaceae Batsch, Tab. Affin. Regni Veg.: 227. 2 Mai 1802, *nom. cons.* Type: *Valeriana* L.

Verbenaceae J. St.-Hil., Expos. Fam. Nat. 1: 245. Feb-Apr 1805, *nom. cons.* Type: *Verbena* L.

Viburnaceae Raf. in Ann. Gén. Sci. Phys. 6: 87. 1820 (to be rejected in favour of *Adoxaceae*). Type: *Viburnum* L.

Violaceae Batsch, Tab. Affin. Regni Veg.: 57. 2 Mai 1802, *nom. cons.* Type: *Viola* L.

Viscaceae Batsch, Tab. Affin. Regni Veg.: 240. 2 Mai 1802 (to be rejected in favour of *Loranthaceae*). Type: *Viscum* L.

Vitaceae Juss., Gen. Pl.: 267. 4 Aug 1789, *nom. cons.* Type: *Vitis* L.

Vivianiaceae Klotzsch in Linnaea 10: 433. Mai 1836. Type: *Viviania* Cav.

Vochysiaceae A. St.-Hil. in Mém. Mus. Hist. Nat. 6: 265. 1820, *nom. cons.* Type: *Vochysia* Aubl.

Wellstediaceae (Pilg.) Novák in Prát, Rostlinopis 9: 530. 1943. Basionym: subfam. *Wellstedioideae* Pilg. in Bot. Jahrb. Syst. 46: 558. 26 Mar 1912. Type: *Wellstedia* Balf. f.

Winteraceae R. Br. ex Lindl., Intr. Nat. Syst. Bot.: 26. Sep 1830, *nom. cons.* Type: *Wintera* Murray, nom. illeg. *(Drimys* J. R. Forst. & G. Forst., *nom. cons.)*

Xanthophyllaceae (Chodat) Gagnep. ex Reveal & Hoogland in Bull. Mus. Natl. Hist. Nat., B, Adansonia 12: 206. 24 Nov 1990. Basionym: trib. *Xanthophylleae* Chodat in Engler & Prantl, Nat. Pflanzenfam. 3(4): 329, 343. Jul 1896. Type: *Xanthophyllum* Roxb., *nom. cons.*

Zygophyllaceae R. Br. in Flinders, Voy. Terra Austr. 2: 545. 19 Jul 1814, *nom. cons.* Type: *Zygophyllum* L.

Monocotyledones

Abolbodaceae Nakai, Chosakuronbun Mokuroku [Ord. Fam. Trib. Nov.]: 221. 20 Jul 1943. Type: *Abolboda* Humb.

Acanthochlamydaceae (Shou C. Chen) P. C. Kao in Acta Bot. Sichuanica 2: 1. 1989. Basionym: subfam. *Acanthochlamydoideae* Shou C. Chen in Acta Phytotax. Sin. 19: 323. 1981. Type: *Acanthochlamys* P. C. Kao

Acoraceae Martinov, Tekhno-Bot. Slovar: 6. 1820. Type: *Acorus* L.

Agavaceae Endl., Ench. Bot.: 105. 15-21 Aug 1841, *nom. cons.* Type: *Agave* L.

Alismataceae Vent., Tabl. Règne Vég. 2: 157. 5 Mai 1799, *nom. cons.* Type: *Alisma* L.

Alliaceae J. Agardh, Theoria Syst. Pl.: 32. Apr-Sep 1858, *nom. cons.* Type: *Allium* L.

Aloaceae Batsch, Tab. Affin. Regni Veg.: 138. 2 Mai 1802. Type: *Aloë* L.

Alstroemeriaceae Dumort., Anal. Fam. Pl.: 57, 58. 1829, *nom. cons.* Type: *Alstroemeria* L.

Amaryllidaceae J. St.-Hil., Expos. Fam. Nat. 1: 134. Feb-Apr 1805, *nom. cons.* Type: *Amaryllis* L.

Anarthriaceae D. F. Cutler & Airy Shaw in Kew Bull. 19: 489. 26 Jul 1965. Type: *Anarthria* R. Br.

Anomochloaceae Nakai, Chosakuronbun Mokuroku [Ord. Fam. Trib. Nov.]: 222. 20 Jul 1943. Type: *Anomochloa* Brongn.

Anthericaceae J. Agardh, Theoria Syst. Pl.: 27. Apr-Sep 1858. Type: *Anthericum* L.

Aphyllanthaceae Burnett, Outl. Bot.: 421, 1153. Jun 1835. Type: *Aphyllanthes* L.

Aponogetonaceae J. Agardh, Theoria Syst. Pl.: 44. Apr-Sep 1858, *nom. cons.* Type: *Aponogeton* L. f., *nom. cons.*

Apostasiaceae Lindl., Nix. Pl.: 22. 17 Sep 1833, *nom. cons.* Type: *Apostasia* Blume

Araceae Juss., Gen. Pl.: 23. 4 Aug 1789, *nom. cons.* Type: *Arum* L.

Arecaceae Schultz Sch., Nat. Syst. Pflanzenr.: 317. 1832, *nom. cons.;* nom. alt.: *Palmae.* Type: *Areca* L.

Asparagaceae Juss., Gen. Pl.: 40. 4 Aug 1789, *nom. cons.* (*Liliaceae* to be treated as having priority). Type: *Asparagus* L.

Asphodelaceae Juss., Gen. Pl.: 51. 4 Aug 1789 (*Liliaceae* to be treated as having priority). Type: *Asphodelus* L.

Asteliaceae Dumort., Anal. Fam. Pl.: 59, 61. 1829 (to be rejected in favour of *Dracaenaceae*). Type: *Astelia* Banks & Sol. ex R. Br., *nom. cons.*

Bambusaceae Burnett, Outl. Bot.: 375. Jun 1835. Type: *Bambusa* Schreb., *nom. cons.*

Blandfordiaceae R. Dahlgren & Clifford in Dahlgren & al., Fam. Monocot.: 155. 1985. Type: *Blandfordia* Sm., *nom. cons.*

Bromeliaceae Juss., Gen. Pl.: 49. 4 Aug 1789, *nom. cons.* Type: *Bromelia* L.

Burmanniaceae Blume, Enum. Pl. Javae: 27. Oct-Dec 1827, *nom. cons.* Type: *Burmannia* L.

Butomaceae Rich. in Mém. Mus. Hist. Nat. 1: 364, 366, 372. 1815-1816, *nom. cons.* Type: *Butomus* L.

Calectasiaceae Endl., Gen. Pl.: viii. Jun 1838. Type: *Calectasia* R. Br.

Calochortaceae Dumort., Anal. Fam. Pl.: 53. 1829. Type: *Calochortus* Pursh

Campynemataceae Dumort., Anal. Fam. Pl.: 57, 58. 1829. Type: *Campynema* Labill.

Cannaceae Juss., Gen. Pl.: 62. 4 Aug 1789, *nom. cons.* Type: *Canna* L.

Cartonemataceae Pichon in Notul. Syst. (Paris) 12: 219. Feb 1946, *nom. cons.* Type: *Cartonema* R. Br.

Centrolepidaceae Endl., Gen. Pl.: 119. Dec 1836, *nom. cons.* Type: *Centrolepis* Labill.

Colchicaceae DC. in Lamarck & Candolle, Fl. Franç., ed. 3, 3: 192. 17 Sep 1805, *nom. cons.* Type: *Colchicum* L.

Commelinaceae R. Br., Prodr.: 268. 27 Mar 1810, *nom. cons.* Type: *Commelina* L.

Conostylidaceae (Pax) Takht., Sist. Magnoliof. [Syst. Magnolioph.]: 313. 31 Mar - 1 Jul 1987. Basionym: trib. *Conostylideae* Pax in Engler & Prantl, Nat. Pflanzenfam. 2(5): 103, 122. Sep 1887. Type: *Conostylis* R. Br.

Convallariaceae Horan., Prim. Lin. Syst. Nat.: 53. Jul-Dec 1834. Type: *Convallaria* L.

Corsiaceae Becc. in Malesia 1: 238. Sep 1878, *nom. cons.* Type: *Corsia* Becc.

Costaceae (Meisn.) Nakai in J. Jap. Bot. 17: 203. Apr 1941. Basionym: trib. *Costeae* Meisn., Pl. Vasc. Gen., Tab. Diagn.: 389; Comm.: 291. 17-20 Aug 1842. Type: *Costus* L.

Croomiaceae Nakai, Iconogr. Pl. As. Orient. 2: 159, t. 60. Nov 1937. Type: *Croomia* Torr.

Cyanastraceae Engl. in Bot. Jahrb. Syst. 28: 357. 22 Mai 1900, *nom. cons.* Type: *Cyanastrum* Oliv.

Cyclanthaceae Dumort., Anal. Fam. Pl.: 65, 66. 1829, *nom. cons.* Type: *Cyclanthus* Poit. ex A. Rich.

Cymodoceaceae N. Taylor in N. Amer. Fl. 17: 31. 30 Jun 1909, *nom. cons.* Type: *Cymodocea* K. D. Koenig, *nom. cons.*

Cyperaceae Juss., Gen. Pl.: 26. 4 Aug 1789, *nom. cons.* Type: *Cyperus* L.

Cypripediaceae Lindl., Nix. Pl.: 22. 17 Sep 1833. Type: *Cypripedium* L.

Dasypogonaceae Dumort., Anal. Fam. Pl.: 54, 55. 1829 (*Xanthorrhoeaceae* to be treated as having priority). Type: *Dasypogon* R. Br.

Dianellaceae Salisb., Gen. Pl.: 66. Apr-Mai 1866. Type: *Dianella* Lam. ex Juss.

Dioscoreaceae R. Br., Prodr.: 294. 27 Mar 1810, *nom. cons.* Type: *Dioscorea* L.

Doryanthaceae R. Dahlgren & Clifford in Dahlgren & al., Fam. Monocot.: 175. 1985. Type: *Doryanthes* Corrêa

Dracaenaceae Salisb., Gen. Pl.: 73. Apr-Mai 1866, *nom. cons.* (to be conserved against *Asteliaceae*). Type: *Dracaena* Vand. ex L.

Ecdeiocoleaceae D. F. Cutler & Airy Shaw in Kew Bull. 19: 495. 26 Jul 1965. Type: *Ecdeiocolea* F. Muell.

Eriocaulaceae P. Beauv. ex Desv. in Ann. Sci. Nat. (Paris) 13: 47. Jan 1828, *nom. cons.* Type: *Eriocaulon* L.

Eriospermaceae Endl., Ench. Bot.: 90. 15-21 Aug 1841. Type: *Eriospermum* Jacq. ex Willd.

Flagellariaceae Dumort., Anal. Fam. Pl.: 59, 60. 1829, *nom. cons.* Type: *Flagellaria* L.

Funkiaceae – see *Hostaceae*

Geosiridaceae Jonker in Receuil Trav. Bot. Néerl. 36 [Meded. Bot. Mus. Herb. Rijks Univ. Utrecht 60]: 477. [seors. impr.] 18 Mai 1939, *nom. cons.* Type: *Geosiris* Baill.

Gramineae Juss., Gen. Pl.: 28. 4 Aug 1789, *nom. cons.;* nom. alt.: *Poaceae*. Type: *Poa* L.

Haemodoraceae R. Br., Prodr.: 299. 27 Mar 1810, *nom. cons.* Type: *Haemodorum* Sm.

Hanguanaceae Airy Shaw in Kew Bull. 18: 260. 8 Dec 1965. Type: *Hanguana* Blume

Heliconiaceae (A. Rich.) Nakai in J. Jap. Bot. 17: 201. Apr 1941. Basionym: trib. *Heliconieae* A. Rich. in Nova Acta Phys.-Med. Acad. Caes. Leop.-Carol. Nat. Cur. 15, Suppl. 13. 1831. Type: *Heliconia* L., *nom. cons.*

Hemerocallidaceae R. Br., Prodr.: 295. 27 Mar 1810. Type: *Hemerocallis* L.

Herreriaceae Endl., Ench. Bot.: 90. 15-21 Aug 1841. Type: *Herreria* Ruiz & Pav.

Hesperocallidaceae Traub in Pl. Life 28: 131. Jan-Apr 1972. Type: *Hesperocallis* A. Gray

Hostaceae B. Mathew in Kew Bull. 43: 302. Jun 1988, *nom. cons. prop.* Type: *Hosta* Tratt., *nom. cons.*

Hyacinthaceae Batsch ex Borkh., Bot. Wörterbuch 1: 315. 1797. Type: *Hyacinthus* L.

Hydatellaceae U. Hamann in New Zealand J. Bot. 14: 195. 22 Jun 1976. Type: *Hydatella* Diels

Hydrocharitaceae Juss., Gen. Pl.: 67. 4 Aug 1789, *nom. cons.* Type: *Hydrocharis* L.

Hypoxidaceae R. Br. in Flinders, Voy. Terra Austr. 2: 576. 19 Jul 1814, *nom. cons.* Type: *Hypoxis* L.

Iridaceae Juss., Gen. Pl.: 57. 4 Aug 1789, *nom. cons.* Type: *Iris* L.

Ixioliriaceae (Pax) Nakai, Chosakuronbun Mokuroku [Ord. Fam. Trib. Nov.]: 234. 20 Jul 1943. Basionym: subtrib. *Ixioliriinae* Pax in Engler & Prantl, Nat. Pflanzenfam. 2(5): 102, 109. Sep 1887. Type: *Ixiolirion* Herb.

Joinvilleaceae Toml. & A. C. Sm. in Taxon 19: 888. 30 Dec 1970. Type: *Joinvillea* Gaudich. ex Brongn. & Gris

Juncaceae Juss., Gen. Pl.: 43. 4 Aug 1789, *nom. cons.* Type: *Juncus* L.

Juncaginaceae Rich., Démonstr. Bot.: ix. Mai 1808, *nom. cons.* (rejected in favour of *Potamogetonaceae*). Type: *Juncago* Hill (*Triglochin* L.)

Lacandoniaceae E. Martínez & Ramos in Ann. Missouri Bot. Gard. 76: 128. 25 Apr 1989. Type: *Lacandonia* E. Martínez & Ramos

Lanariaceae H. Huber ex R. Dahlgren in Monogr. Syst. Bot. Missouri Bot. Gard. 25: 71. Jun 1988. Type: *Lanaria* Aiton, *nom. cons.*

Lemnaceae Gray, Nat. Arr. Brit. Pl. 2: 729. 1 Nov 1821, *nom. cons.* Type: *Lemna* L.

Lilaeaceae Dumort., Anal. Fam. Pl.: 62, 65. 1829, *nom. cons.* Type: *Lilaea* Bonpl.

Liliaceae Juss., Gen. Pl.: 48. 4 Aug 1789, *nom. cons.* (to be treated as having priority over *Asparagaceae* and *Asphodelaceae*). Type: *Lilium* L.

Limnocharitaceae Takht. ex Cronquist, Integr. Syst. Classif. Fl. Pl.: 1048. 1981. Type: *Limnocharis* Bonpl.

Lomandraceae Lotsy, Vortr. Bot. Stammesgesch. 3: 761. Sep-Oct 1911. Type: *Lomandra* Labill.

Lowiaceae Ridl., Fl. Malay Penins. 4: 291. 1 Dec 1924, *nom. cons.* Type: *Lowia* Scort. [= *Orchidantha* N. E. Br.]

Luzuriagaceae Lotsy, Vortr. Bot. Stammesgesch. 3: 760. Sep-Oct 1911. Type: *Luzuriaga* Ruiz & Pav., *nom. cons.*

Marantaceae Petersen in Engler & Prantl, Nat. Pflanzenfam. 2(6): 33. Oct 1888, *nom. cons.* Type: *Maranta* L.

Maundiaceae Nakai, Chosakuronbun Mokuroku [Ord. Fam. Trib. Nov.]: 213. 20 Jul 1943. Type: *Maundia* F. Muell.

Mayacaceae Kunth in Abh. Königl. Akad. Wiss. Berlin, Phys. Abh. 1840: 93. 1842, *nom. cons.* Type: *Mayaca* Aubl.

Medeolaceae (S. Watson) Takht., Sist. Magnoliof. [Syst. Magnolioph.]: 291. 31 Mar - 1 Jul 1987. Basionym: subtrib. *Medeolinae* S. Watson in Proc. Amer. Acad. Arts 14: 222. 1879. Type: *Medeola* L.

Melanthiaceae Batsch, Tab. Affin. Regni Veg.: 133. 2 Mai 1802, *nom. cons.* Type: *Melanthium* L.

Milulaceae Traub in Pl. Life 28: 131. Jan-Apr 1972. Type: *Milula* Prain

Musaceae Juss., Gen. Pl.: 61. 4 Aug 1789, *nom. cons.* Type: *Musa* L.

Najadaceae Juss., Gen. Pl.: 18. 4 Aug 1789, *nom. cons.* Type: *Najas* L.

Neuwiediaceae (Burns-Bal. & V. A. Funk) R. Dahlgren ex Reveal & Hoogland in Bull. Mus. Natl. Hist. Nat., B, Adansonia 13: 91. 4 Oct 1991. Basionym: subfam. *Neuwiedioideae* Burns-Bal. & V. A. Funk in Smithsonian Contr. Bot. 61: 53. 26 Mar 1986. Type: *Neuwiedia* Blume

Nolinaceae Nakai, Chosakuronbun Mokuroku [Ord. Fam. Trib. Nov.]: 226. 20 Jul 1943. Type: *Nolina* Michx.

Nypaceae Brongn. ex Le Maout & Decne., Traite Général Bot.: 624. Jan-Apr 1868. Type: *Nypa* Steck

Ophiopogonaceae Endl., Ench. Bot.: 90. 15-21 Aug 1841. Type: *Ophiopogon* Ker Gawl., *nom. cons.*

Orchidaceae Juss., Gen. Pl.: 64. 4 Aug 1789, *nom. cons.* Type: *Orchis* L.

Palmae Juss., Gen. Pl.: 37. 4 Aug 1789, *nom. cons.;* nom. alt.: *Arecaceae.* Type: *Areca* L.

Pandanaceae R. Br., Prodr.: 340. 27 Mar 1810, *nom. cons.* Type: *Pandanus* Parkinson

Petermanniaceae Hutch., Fam. Fl. Pl. 2: 113. 1934, *nom. cons.* Type: *Petermannia* F. Muell., *nom. cons.*

Petrosaviaceae Hutch., Fam. Fl. Pl. 2: 36. 1934, *nom. cons.* Type: *Petrosavia* Becc.

Philesiaceae Dumort., Anal. Fam. Pl.: 53, 54, 97. 1829, *nom. cons.* Type: *Philesia* Comm. ex Juss.

Philydraceae Link, Enum. Hort. Berol. Alt. 1: 5. 16 Mar - 30 Jun 1821, *nom. cons.* Type: *Philydrum* Banks ex Sol.

Phormiaceae J. Agardh, Theoria Syst. Pl.: 7. Apr-Sep 1858. Type: *Phormium* J. R. Forst. & G. Forst.

Poaceae (R. Br.) Barnhart in Bull. Torrey Bot. Club 22: 7. 15 Jan 1895, *nom. cons.;* nom. alt.: *Gramineae.* Basionym: trib. *Poeae* R. Br. in Flinders, Voy. Terra Austr. 2: 583. 19 Jul 1814. Type: *Poa* L.

Pontederiaceae Kunth in Humboldt & al., Nov. Gen. Sp. 1, ed. 4°: 265. Mai 1816, *nom. cons.* Type: *Pontederia* L.

Posidoniaceae Lotsy, Vortr. Bot. Stammesgesch. 3: 658. Sep-Oct. 1911, *nom. cons.* Type: *Posidonia* K. D. Koenig, *nom. cons.*

Potamogetonaceae Dumort., Anal. Fam. Pl.: 59, 61. 1829, *nom. cons.* (conserved against *Juncaginaceae*). Type: *Potamogeton* L.

Rapateaceae Dumort., Anal. Fam. Pl.: 60, 62. 1829, *nom. cons.* Type: *Rapatea* Aubl.

Restionaceae R. Br., Prodr.: 243. 27 Mar 1810, *nom. cons.* Type: *Restio* Rottb., *nom. cons.*

Rhipogonaceae Conran & Clifford in Nordic J. Bot. 5: 219. 25 Oct 1985. Type: *Rhipogonum* J. R. Forst. & G. Forst.

Roxburghiaceae – see *Stemonaceae.*

Ruppiaceae Horan. ex Hutch., Fam. Fl. Pl. 2: 48. 1934, *nom. cons.* Type: *Ruppia* L.

Ruscaceae Spreng. ex Hutch., Fam. Fl. Pl. 2: 109. 1934, *nom. cons.* Type: *Ruscus* L.

Scheuchzeriaceae F. Rudolphi, Syst. Orb. Veg.: 28. 1830, *nom. cons.* Type: *Scheuchzeria* L.

Smilacaceae Vent., Tabl. Règne Vég. 2: 146. 5 Mai 1799, *nom. cons.* Type: *Smilax* L.

Sparganiaceae F. Rudolphi, Syst. Orb. Veg.: 27. 1830, *nom. cons.* Type: *Sparganium* L.

Stemonaceae Engl. in Engler & Prantl, Nat. Pflanzenfam. 2(5): 8. 26 Mar 1887, *nom. cons.* Type: *Stemona* Lour.

Stenomeridaceae J. Agardh, Theoria Syst. Pl.: 66. Apr-Sep 1858, *nom. cons.* Type: *Stenomeris* Planch.

Strelitziaceae (K. Schum.) Hutch., Fam. Fl. Pl. 2: 72. 1934, *nom. cons.* Basionym: subfam. *Strelitzioideae* K. Schum. in Engler, Pflanzenr. 1: 13. 4 Oct 1900. Type: *Strelitzia* Aiton

Streptochaetaceae Nakai, Chosakuronbun Mokuroku [Ord. Fam. Trib. Nov.]: 222. 20 Jul 1943. Type: *Streptochaeta* Schrad. ex Nees

Taccaceae Dumort., Anal. Fam. Pl.: 57, 58. 1829, *nom. cons.* Type: *Tacca* J. R. Forst. & G. Forst., *nom. cons.*

Tecophilaeaceae Leyb. in Bonplandia 10: 370. 1862, *nom. cons.* Type: *Tecophilaea* Bertero ex Colla

Thalassiaceae Nakai, Chosakuronbun Mokuroku [Ord. Fam. Trib. Nov.]: 210. 20 Jul 1943. Type: *Thalassia* Banks & Sol. ex K. D. Koenig

Thismiaceae J. Agardh, Theoria Syst. Pl.: 99. Apr-Sep 1858, *nom. cons.* Type: *Thismia* Griff.

Thurniaceae Engl., Syllabus, ed. 5: 94. Jul 1907, *nom. cons.* Type: *Thurnia* Hook. f.

Trichopodaceae Hutch., Fam. Fl. Pl. 2: 143. 1934, *nom. cons.* Type: *Trichopus* Gaertn.

Tricyrtidaceae – see *Uvulariaceae*

Trilliaceae Lindl., Veg. Kingd.: 218. Jan-Mai 1846, *nom. cons.* Type: *Trillium* L.

Triuridaceae Gardner in Trans. Linn. Soc. London 19: 160. Jun-Dec 1843, *nom. cons.* Type: *Triuris* Miers

Typhaceae Juss., Gen. Pl.: 25. 4 Aug 1789, *nom. cons.* Type: *Typha* L.

Uvulariaceae A. Gray ex Kunth, Enum. Pl. 4: 199. 17-19 Jul 1843, *nom. cons. prop.* Type: *Uvularia* L.

Velloziaceae Endl., Ench. Bot.: 101. 15-21 Aug 1841, *nom. cons.* Type: *Vellozia* Vand.

Xanthorrhoeaceae Dumort., Anal. Fam. Pl.: 60, 62, 103. 1829, *nom. cons.* (to be treated as having priority over *Dasypogonaceae*). Type: *Xanthorrhoea* Sm., *nom. cons. prop.*

Xyridaceae C. Agardh, Aphor. Bot.: 158. 23 Mai 1823, *nom. cons.* Type: *Xyris* L.

Zannichelliaceae Dumort., Anal. Fam. Pl.: 59, 61. 1829, *nom. cons.* Type: *Zannichellia* L.

Zingiberaceae Lindl., Key Bot.: 69. 15-30 Sep 1835, *nom. cons.* Type: *Zingiber* Boehm., *nom. cons.*

Zosteraceae Dumort., Anal. Fam. Pl.: 65, 66. 1829, *nom. cons.* Type: *Zostera* L.

BRYOPHYTE FAMILY NAMES IN CURRENT USE

By Marshall J. Crosby (Musci),

Riclef Grolle and Gea Zijlstra (Hepaticae)

Selection of the bryophyte family names listed below was under the responsibility of Crosby for the Musci and of Grolle for the Hepaticae. Illegitimate names were of course omitted, and so were legitimate names that are straightforward synonyms of earlier legitimate names. Even so, many recently segregated monogeneric or even monospecific families had to be taken up, which will hardly ever come into general use.

The Committee for Bryophyta was consulted with respect to a number of questions related to this list, and proved to be very helpful. We want, in particular, to acknowledge specific comments received from Hélène Bischler, Rob S. Gradstein, Pekka Isoviita, and Jiří Váňa.

Musci

Adelotheciaceae W. R. Buck in Brittonia 39: 214. 1987. Type: *Adelothecium* Mitt.

Amblystegiaceae G. Roth in Hedwigia 38: (6). 1899. Type: *Amblystegium* Schimp.

Andreaeaceae Dumort., Anal. Fam. Pl.: 68. 1829. Type: *Andreaea* Hedw.

Andreaeobryaceae Steere in Phytologia 33: 410. 1976. Type: *Andreaeobryum* Steere & B. M. Murray

Archidiaceae Schimp., Coroll. Bryol. Eur.: 5. 1856. Type: *Archidium* Brid.

Aulacomniaceae Schimp., Syn. Musc. Eur.: 411. Mar-Apr 1860. Type: *Aulacomnium* Schwägr., *nom. cons.*

Bartramiaceae Schwägr. in Willdenow, Sp. Pl. 5(2): 90. 1830. Type: *Bartramia* Hedw., *nom. cons.*

Brachytheciaceae Schimp., Syn. Musc. Eur., ed. 2: "xcv" [cxv]. Apr 1876. Type: *Brachythecium* Schimp.

Bruchiaceae Schimp., Coroll. Bryol. Eur.: 6. 1856. Type: *Bruchia* Schwägr.

Bryaceae Schwägr. in Willdenow, Sp. Pl. 5(2): 47. 1830. Type: *Bryum* Hedw.

Bryobartramiaceae Sainsbury in Bryologist 51: 9. 1948. Type: *Bryobartramia* Sainsbury

Bryoxiphiaceae Besch. in J. Bot. (Morot) 6: 183. 1892, *nom. cons.* Type: *Bryoxiphium* Mitt., *nom. cons.*

Buxbaumiaceae Schwägr. in Willdenow, Sp. Pl. 5(2): 23. 1830. Type: *Buxbaumia* Hedw.

Callicostaceae H. A. Crum in Bryologist 87: 210. 1984. Type: *Callicosta* Müll. Hal.

Calomniaceae Kindb. in Bot. Centralbl. 77: 394. 1899. Type: *Calomnion* Hook. f. & Wilson

Calymperaceae Kindb., Gen. Eur. N.-Amer. Bryin.: 11. 1897. Type: *Calymperes* Sw. ex F. Weber

Catagoniaceae W. R. Buck & Ireland in Nova Hedwigia 41: 115. 1985. Type: *Catagonium* Müll. Hal. ex Broth.

Catoscopiaceae Boulay ex Broth. in Engler & Prantl, Nat. Pflanzenfam. 1(3): 629. 18 Oct 1904. Type: *Catoscopium* Brid.

Climaciaceae Kindb., Gen. Eur. N.-Amer. Bryin.: 7. 1897. Type: *Climacium* F. Weber & D. Mohr

Cochlearidiaceae Ochyra & Mahu in J. Bryol. 16: 221. 1990. Type: *Cochlearidium* Ochyra & Mahu

Cryphaeaceae Schimp., Coroll. Bryol. Eur.: 97. 1856. Type: *Cryphaea* F. Weber

Cyrtopodaceae M. Fleisch., Musc. Buitenzorg: 655. Oct-Dec 1908. Type: *Cyrtopus* (Brid.) Hook. f.

Daltoniaceae Schimp., Syn. Musc. Eur.: 478. Mar-Apr 1860. Type: *Daltonia* Hook. & Taylor, *nom. cons.*

Dawsoniaceae Broth. in Engler & Prantl, Nat. Pflanzenfam. 1(3): 698. 7 Mar 1905. Type: *Dawsonia* R. Br.

Dicnemonaceae Broth. in Engler & Prantl, Nat. Pflanzenfam., ed. 2, 10: 214. Mai-Jun 1924. Type: *Dicnemon* Schwägr.

Dicranaceae Schimp., Coroll. Bryol. Eur.: 11. 1856. Type: *Dicranum* Hedw.

Diphysciaceae M. Fleisch. in Engler, Syllabus, ed. 8: 88. Jan-Feb 1920. Type: *Diphyscium* D. Mohr

Disceliaceae Schimp., Coroll. Bryol. Eur.: 59. 1856. Type: *Discelium* Brid.

Distichophyllaceae H. A. Mill. in Phytologia 21: 249. 1971. Type: *Distichophyllum* Dozy & Molk.

Ditrichaceae Limpr., Laubm. Deutschl. 1: 482. Oct 1887, *nom. cons.* Type: *Ditrichum* Hampe, *nom. cons.*

Donrichardsiaceae Ochyra in Lindbergia 11: 7. 1985. Type: *Donrichardsia* H. A. Crum & L. E. Anderson

Echinodiaceae Broth. in Engler & Prantl, Nat. Pflanzenfam. 1(3): 1216. 16 Mar 1909. Type: *Echinodium* Jur.

Encalyptaceae Schimp., Coroll. Bryol. Eur.: 38. 1856. Type: *Encalypta* Hedw.

Entodontaceae Kindb., Gen. Eur. N.-Amer. Bryin.: 7. 1897, *nom. cons.* Type: *Entodon* Müll. Hal.

Ephemeraceae Schimp., Coroll. Bryol. Eur.: 3. 1856. Type: *Ephemerum* Hampe, *nom. cons.*

Ephemeropsidaceae W. Schultze-Motel in Taxon 19: 251. 29 Apr

1970. Type: *Ephemeropsis* K. I. Goebel

Erpodiaceae Broth. in Engler & Prantl, Nat. Pflanzenfam. 1(3): 706. 7 Mar 1905. Type: *Erpodium* (Brid.) Brid.

Eustichiaceae Broth. in Engler & Prantl, Nat. Pflanzenfam., ed. 2, 10: 420. Mai-Jun 1924, *nom. cons.* Type: *Eustichia* Mitt.

Fabroniaceae Schimp., Coroll. Bryol. Eur.: 102. 1856. Type: *Fabronia* Raddi

Fissidentaceae Schimp., Coroll. Bryol. Eur.: 96. 1856. Type: *Fissidens* Hedw.

Fontinalaceae Schimp., Coroll. Bryol. Eur.: 96. 1856. Type: *Fontinalis* Hedw.

Funariaceae Schwägr. in Willdenow, Sp. Pl. 5(2): 43. 1830. Type: *Funaria* Hedw.

Gigaspermaceae Lindb. in Ofvers. Förh. Kongl. Svenska Vetensk.-Acad. 1864: 592. 1865. Type: *Gigaspermum* Lindb.

Grimmiaceae Arn., Disp. Méth. Mousses: 19. 20 Dec 1825 - 6 Feb 1826. Type: *Grimmia* Hedw.

Hedwigiaceae Schimp., Coroll. Bryol. Eur.: 52. 1856. Type: *Hedwigia* P. Beauv., *nom. cons.*

Helicophyllaceae Broth. in Engler & Prantl, Nat. Pflanzenfam. 1(3): 973. 1 Oct 1907. Type: *Helicophyllum* Brid.

Hookeriaceae Schimp., Coroll. Bryol. Eur.: 101. 1856. Type: *Hookeria* Sm., *nom. cons.*

Hydropogonaceae W. H. Welch in Bryologist 46: 31. 1943. Type: *Hydropogon* Brid.

Hylocomiaceae (Broth.) M. Fleisch. in Nova Guinea 12: 125. 1914. Basionym: subfam. *Hylocomioideae* Broth. in Engler & Prantl, Nat. Pflanzenfam. 1(3): 1044. 23 Jun 1908. Type: *Hylocomium* Schimp.

Hypnaceae Schimp., Coroll. Bryol. Eur.: 113. 1856. Type: *Hypnum* Hedw., *nom. cons.*

Hypnobartlettiaceae Ochyra in Lindbergia 11: 7. 1985. Type: *Hypnobartlettia* Ochyra

Hypnodendraceae Broth. in Engler & Prantl, Nat. Pflanzenfam. 1(3): 1166. 16 Mar 1909. Type: *Hypnodendron* (Müll. Hal.) Lindb.

Hypopterygiaceae Mitt. in J. Proc. Linn. Soc., Bot., Suppl. 1: 147. 1859. Type: *Hypopterygium* Brid.

Lembophyllaceae Broth. in Engler & Prantl, Nat. Pflanzenfam. 1(3): 863. 31 Jul 1906. Type: *Lembophyllum* Lindb.

Leptodontaceae Schimp., Coroll. Bryol. Eur.: 98. 1856. Type: *Leptodon* D. Mohr, *nom. cons.*

Leptostomaceae Schwägr. in Willdenow, Sp. Pl. 5(2): 85. 1830. Type: *Leptostomum* R. Br., *nom. cons.*

Lepyrodontaceae Broth. in Engler & Prantl, Nat. Pflanzenfam. 1(3): 771. 19 Apr 1906. Type: *Lepyrodon* Hampe

Lepyrodontopsidaceae W. R. Buck in Brittonia 33: 560. 1981. Type: *Lepyrodontopsis* Broth.

Leskeaceae Schimp., Coroll. Bryol. Eur.: 109. 1856. Type: *Leskea* Hedw.

Leucobryaceae Schimp., Coroll. Bryol. Eur.: 19. 1856. Type: *Leucobryum* Hampe

Leucodontaceae Schimp., Coroll. Bryol. Eur.: 108. 1856. Type: *Leucodon* Schwägr.

Leucomiaceae Broth. in Engler & Prantl, Nat. Pflanzenfam. 1(3): 1095. 10 Nov 1908. Type: *Leucomium* Mitt.

Meesiaceae Schimp., Coroll. Bryol. Eur.: 82. 1856. Type: *Meesia* Hedw., *nom. cons.*

Meteoriaceae Kindb., Gen. Eur. N.-Amer. Bryin.: 7. 1897. Type: *Meteorium* (Brid.) Dozy & Molk.

Microtheciellaceae H. A. Mill. & A. J. Harr. in J. Bryol. 9: 523. 1977. Type: *Microtheciella* Dixon

Mitteniaceae Broth. in Engler & Prantl, Nat. Pflanzenfam. 1(3): 532. 24 Nov 1903. Type: *Mittenia* Lindb.

Mniaceae Schwägr. in Willdenow, Sp. Pl. 5(2): 25. 1830. Type: *Mnium* Hedw., *nom. cons.*

Myuriaceae M. Fleisch., Musc. Buitenzorg: 1663. Feb-Dec 1923. Type: *Myurium* Schimp.

Nanobryaceae W. Schultze-Motel in Willdenowia 5: 386. 1969. Type: *Nanobryum* Dixon

Neckeraceae Schimp., Coroll. Bryol. Eur.: 99. 1856. Type: *Neckera* Hedw., *nom. cons.*

Oedipodiaceae Schimp., Syn. Musc. Eur., ed. 2: xcviii, 354. Apr 1876. Type: *Oedipodium* Schwägr.

Orthorrhynchiaceae S. H. Lin in J. Taiwan Mus. 36(2): 74. 1983. Type: *Orthorrhynchium* Reichardt

Orthotrichaceae Arn., Disp. Méth. Mousses: 13. 20 Dec 1825 - 6 Feb 1826. Type: *Orthotrichum* Hedw.

Phyllodrepaniaceae Crosby in Rev. Bryol. Lichénol. 37: 347. 1971. Type: *Phyllodrepanium* Crosby

Phyllogoniaceae Kindb. in Bot. Centralbl. 76: 85. 1898. Type: *Phyllogonium* Brid.

Plagiotheciaceae (Broth.) M. Fleisch. in Nova Guinea 8: 748. 1912. Basionym: subfam. *Plagiothecioideae* Broth. in Engler & Prantl, Nat. Pflanzenfam. 1(3): 1078. 10 Nov 1908. Type: *Plagiothecium* Schimp.

Pleurophascaceae Broth. in Engler & Prantl, Nat. Pflanzenfam. 1(3): 774. 19 Apr 1906. Type: *Pleurophascum* Lindb.

Pleuroziopsidaceae Ireland in J. Hattori Bot. Lab. 31: 62. 1968. Type: *Pleuroziopsis* Kindb. ex E. Britton

Polytrichaceae Schwägr. in Willdenow, Sp. Pl. 5(2): 1. 1830. Type: *Polytrichum* Hedw.

Pottiaceae Schimp., Coroll. Bryol. Eur.: 24. 1856, *nom. cons.* Type: *Pottia* (Ehrh. ex Rchb.) Fürnr.

Prionodontaceae Broth. in Engler & Prantl, Nat. Pflanzenfam. 1(3): 763. 1 Aug 1905. Type: *Prionodon* Müll. Hal.

Pseudoditrichaceae Steere & Z. Iwats. in Bryologist 78: 490. 1976. Type: *Pseudoditrichum* Steere & Z. Iwats.

Pterigynandraceae Schimp., Syn. Musc. Eur., ed. 2: cxiii, 618. Apr 1876. Type: *Pterigynandrum* Hedw.

Pterobryaceae Kindb., Gen. Eur. N.-Amer. Bryin.: 15. 1897. Type: *Pterobryon* Hornsch.

Ptychomitriaceae Schimp., Syn. Musc. Eur.: 241. Mar-Apr 1860. Type: *Ptychomitrium* Fürnr., *nom. cons.*

Ptychomniaceae M. Fleisch., Musc. Buitenzorg: 656. Oct-Dec 1908. Type: *Ptychomnion* (Hook. f. & Wilson) Mitt.

Racopilaceae Kindb. in Bot. Centralbl. 76: 85. 1898. Type: *Racopilum* P. Beauv.

Regmatodontaceae Broth. in Engler & Prantl, Nat. Pflanzenfam. 1(3): 1125. 10 Nov 1908. Type: *Regmatodon* Brid.

Rhabdoweisiaceae Limpr., Laubm. Deutschl. 1: 271. Nov. 1886. Type: *Rhabdoweisia* Bruch & Schimp.

Rhachitheciaceae H. Rob. in Bryologist 67: 448. 1964. Type: *Rhachithecium* Broth. ex Le Jol.

Rhizogoniaceae Broth. in Engler & Prantl, Nat. Pflanzenfam. 1(3): 614. 15 Jan 1904. Type: *Rhizogonium* Brid.

Rhytidiaceae Broth. in Engler & Prantl, Nat. Pflanzenfam., ed. 2, 11: 475. Mai-Jun 1925. Type: *Rhytidium* (Sull.) Kindb.

Rigodiaceae H. A. Crum in Bryologist 84: 392. 1981. Type: *Rigodium* Kunze ex Schwägr.

Rutenbergiaceae (Broth.) M. Fleisch., Musc. Buitenzorg: 725. Oct-Dec 1908. Basionym: subfam. *Rutenbergioideae* Broth. in Engler & Prantl, Nat. Pflanzenfam. 1(3): 786. 19 Apr 1906. Type: *Rutenbergia* Geh. & Hampe ex Besch.

Schistostegaceae Schimp., Coroll. Bryol. Eur.: 54. 1856. Type: *Schistostega* D. Mohr

Scouleriaceae S. P. Churchill in Funk & Brooks, Advances Cladistics: 143. 1981. Type: *Scouleria* Hook.

Seligeriaceae Schimp., Coroll. Bryol. Eur.: 22. 1856. Type: *Seligeria* Bruch & Schimp.

Sematophyllaceae Broth. in Engler & Prantl, Nat. Pflanzenfam. 1(3): 1098. 10 Nov 1908, *nom. cons.* Type: *Sematophyllum* Mitt.

Sorapillaceae M. Fleisch., Musc. Buitenzorg: 847. Oct-Dec 1908. Type: *Sorapilla* Spruce & Mitt.

Sphagnaceae Dumort., Anal. Fam. Pl.: 68. 1829. Type: *Sphagnum* L.

Spiridentaceae Kindb. in Bot. Centralbl. 76: 85. 1898. Type: *Spiridens* Nees

Splachnaceae Grev. & Arn. in Mem. Wern. Nat. Hist. Soc. 5: 442. 1824. Type: *Splachnum* Hedw.

Splachnobryaceae A. K. Kop. in Ann. Bot. Fenn. 18: 128. 1981. Type: *Splachnobryum* Müll. Hal.

Stereophyllaceae (M. Fleisch.) W. R. Buck & Ireland in Nova Hedwigia 41: 95. 1985. Basionym: subfam. *Stereophylloideae* M. Fleisch., Musc. Buitenzorg: 1158. Feb-Dec 1923. Type: *Stereophyllum* Mitt.

Tetraphidaceae Schimp., Coroll. Bryol. Eur.: 37. 1856. Type: *Tetraphis* Hedw.

Thamnobryaceae Margad. & During, Beknopte Fl. Ned. Blad- & Leverm.: 443. 1982. Type: *Thamnobryum* Nieuwl.

Theliaceae (Broth.) M. Fleisch., Musc. Buitenzorg: 1485. Feb-Dec 1923. Basionym: subfam. *Thelioideae* Broth. in Engler & Prantl, Nat. Pflanzenfam. 1(3): 981. 1 Oct 1907. Type: *Thelia* Sull.

Thuidiaceae Schimp., Syn. Musc. Eur.: 493. Mar-Apr 1860. Type: *Thuidium* Bruch & Schimp.

Timmiaceae Schimp., Coroll. Bryol. Eur.: 87. 1856. Type: *Timmia* Hedw., *nom. cons.*

Trachypodaceae M. Fleisch. in Hedwigia 45: 63. 1905. Type: *Trachypus* Reinw. & Hornsch.

Viridivelleraceae I. G. Stone in J. Bryol. 9: 21. 1976. Type: *Viridivellus* I. G. Stone

Wardiaceae W. H. Welch in Bryologist 46: 26. 1943. Type: *Wardia* Harv. & Hook.

Hepaticae

Acrobolbaceae E. A. Hodgs. in Rec. Domin. Mus. 4: 177. 1962. Type: *Acrobolbus* Nees

Adelanthaceae (Jörg.) Grolle in J. Hattori Bot. Lab. 35: 327. 23 Mar 1972. Basionym: trib. *Adelantheae* Jörg. in Bergens Mus. Skr. 16: 257. 1934. Type: *Adelanthus* Mitt., *nom. cons.*

Allisoniaceae (R. M. Schust. ex Grolle) Schljakov, Pecenocnye Mchi, Morfol. Filog. Klassif.: 119. 1975. Basionym: subfam. *Allisonioideae* R. M. Schust. ex Grolle in J. Bryol. 7: 204. 1972. Type: *Allisonia* Herzog

Aneuraceae H. Klinggr., Höh. Crypt. Preuss.: 11. Apr-Dec 1858. Type: *Aneura* Dumort., *nom. cons. prop.*

Antheliaceae R. M. Schust. in J. Hattori Bot. Lab. 26: 236. 29 Aug 1963. Type: *Anthelia* (Dumort.) Dumort.

Anthocerotaceae Dumort., Anal. Fam. Pl.: 68, 69. 1829. Type: *Anthoceros* L.

Arnelliaceae Nakai, Chosakuronbun Mokuroku [Ord. Fam. Trib. Nov.]: 200. 20 Jul 1943. Type: *Arnellia* Lindb.

Aytoniaceae Cavers in New Phytol. 10: 42. 1911. Type: *Aytonia* J. R. Forst. & G. Forst., *nom. rejic.* [= *Plagiochasma* Lehm. & Lindenb.]

Balantiopsidaceae H. Buch in Mitt. Thüring. Bot. Ges. 1(2-3): 23. 1955. Type: *Balantiopsis* Mitt.

Blasiaceae H. Klinggr., Höh. Crypt. Preuss.: 14. Apr-Dec 1858. Type: *Blasia* L.

Brevianthaceae J. J. Engel & R. M. Schust. in Phytologia 47: 317.

28 Jan 1981. Type: *Brevianthus* J. J. Engel & R. M. Schust.

Calypogeiaceae (Müll. Frib.) Arnell in Holmberg, Skand. Fl. 2a: 189. 20 Jan 1928. Basionym: subfam. *Calypogeioideae* Müll. Frib., Lebermoose 2: 6. 1 Mar 1912. Type: *Calypogeia* Raddi, *nom. cons.*

Cephaloziaceae Mig., Krypt.-Fl. Deutschl., Moose: 465. Jul 1904. Type: *Cephalozia* (Dumort.) Dumort.

Cephaloziellaceae Douin in Bull. Soc. Bot. France, Mém. 29: 1, 5, 13. 1920. Type: *Cephaloziella* (Spruce) Schiffn., *nom. cons.*

Chaetophyllopsidaceae R. M. Schust. in J. Hattori Bot. Lab. 23: 68. 22 Mar 1961. Type: *Chaetophyllopsis* R. M. Schust.

Chonecoleaceae R. M. Schust. ex Grolle in J. Bryol. 7: 206. 1972. Type: *Chonecolea* Grolle

Cleveaceae Cavers in New Phytol. 10: 42. 1911. Type: *Clevea* Lindb. [= *Athalamia* Falc.]

Conocephalaceae Müll. Frib. ex Grolle in J. Bryol. 7: 207. 1972. Type: *Conocephalum* Hill, *nom. cons.*

Corsiniaceae Engl., Syllabus, Grosse Ausgabe: 44. Apr 1892. Type: *Corsinia* Raddi

Dendrocerotaceae (Milde) Hässel in J. Hattori Bot. Lab. 64: 82. 14 Jun 1988. Basionym: subtrib. *Dendrocerotinae* Milde in Bot. Zeitung (Berlin) 17: 54. 11 Feb 1859. Type: *Dendroceros* Nees

Exormothecaceae Müll. Frib. ex Grolle in J. Bryol. 7: 208. 1972. Type: *Exormotheca* Mitt.

Fossombroniaceae Hazsl., Magyar Birodalom Moh-Flórája: 20, 36. Mai-Dec 1885, *nom. cons. prop.* Type: *Fossombronia* Raddi

Frullaniaceae Lorch in Lindau, Krypt.-Fl. Anf. 6: 174. Jan 1914. Type: *Frullania* Raddi

Geocalycaceae H. Klinggr., Höh. Crypt. Preuss.: 34. Apr-Dec 1858. Type: *Geocalyx* Nees

Goebeliellaceae Verd., Man. Bryol.: 425. Sep 1932. Type: *Goebeliella* Steph.

Grolleaceae Solari ex R. M. Schust. in Phytologia 56: 66. 11 Aug 1984. Type: *Grollea* R. M. Schust.

Gymnomitriaceae H. Klinggr., Höh. Crypt. Preuss.: 16. Apr-Dec 1858. Type: *Gymnomitrion* Corda, *nom. cons.*

Gyrothyraceae R. M. Schust. in Trans. Brit. Bryol. Soc. 6: 87. 1970. Type: *Gyrothyra* M. Howe

Haplomitriaceae Dedecek in Arch. Přír. Proskoumání Čech 5(4): 71. 1884. Type: *Haplomitrium* Nees, *nom. cons.*

Herbertaceae Müll. Frib. ex Fulford & Hatcher in Bryologist 61: 284. Dec 1958. Type: *Herbertus* Gray

Hymenophytaceae R. M. Schust. in J. Hattori Bot. Lab. 26: 296. 29 Aug 1963. Type: *Hymenophyton* Dumort.

Jackiellaceae R. M. Schust. in J. Hattori Bot. Lab. 36: 395. 5 Jan 1973. Type: *Jackiella* Schiffn.

Jubulaceae H. Klinggr., Höh. Crypt. Preuss.: 40. Apr-Dec 1858. Type: *Jubula* Dumort., *nom. cons. prop.*

Jubulopsidaceae (Hamlin) R. M. Schust. in Phytologia 56: 68. 11 Aug 1984. Basionym: subfam. *Jubulopsidoideae* Hamlin in J. Hattori Bot. Lab. 37: 176. 1973. Type: *Jubulopsis* R. M. Schust.

Jungermanniaceae Rchb., Bot. Damen: 256. 1828. Type: *Jungermannia* L.

Leiosporocerotaceae Hässel in J. Bryol. 14: 255. 16 Dec 1986. Type: *Leiosporoceros* Hässel

Lejeuneaceae Casares-Gil, Fl. Ibér. Briόf. 1: 703. Dec 1919, *nom. cons.* Type: *Lejeunea* Lib., *nom. cons.*

Lepicoleaceae R. M. Schust. in Nova Hedwigia 5: 27. 31 Jan 1963. Type: *Lepicolea* Dumort.

Lepidolaenaceae Nakai, Chosakuronbun Mokuroku [Ord. Fam. Trib. Nov.]: 200. 20 Jul 1943. Type: *Lepidolaena* Dumort.

Lepidoziaceae Limpr. in Cohn, Krypt.-Fl. Schlesien 1: 310. Mai 1877. Type: *Lepidozia* (Dumort.) Dumort., *nom. cons.*

Lophocoleaceae (Jörg.) Vanden Berghen in Robyns, Fl. Gén. Belgique, Bryophytes 1: 208. 1956. Basionym: trib. *Lophocoleeae* Jörg. in Bergens Mus. Skr. 16: 61. 1934. Type: *Lophocolea* (Dumort.) Dumort.

Lophoziaceae Cavers in New Phytol. 9: 293. 1910. Type: *Lophozia* (Dumort.) Dumort.

Lunulariaceae H. Klinggr., Höh. Crypt. Preuss.: 9. Apr-Dec 1858. Type: *Lunularia* Adans.

Makinoaceae Nakai, Chosakuronbun Mokuroku [Ord. Fam. Trib. Nov.]: 201. 20 Jul 1943. Type: *Makinoa* Miyake

Marchantiaceae (Bisch.) Lindl., Nat. Syst. Bot., ed. 2: xxvi, 412. Jul 1836. Basionym: trib. *Marchantieae* Bisch., Bem. Leberm.: 53. 1835. Type: *Marchantia* L.

Mesoptychiaceae Inoue & Steere in Bull. Natl. Sci. Mus., Tokyo, B, 1: 62. 22 Jun 1975. Type: *Mesoptychia* (Lindb.) A. Evans

Metzgeriaceae H. Klinggr., Höh. Crypt. Preuss.: 10. Apr-Dec 1858. Type: *Metzgeria* Raddi

Mizutaniaceae Furuki & Z. Iwats. in J. Hattori Bot. Lab. 67: 291. 26 Dec 1989. Type: *Mizutania* Furuki & Z. Iwats.

Monocarpaceae D. J. Carr ex Schelpe in J. S. African Bot. 35: 110. 15 Mar 1969. Type: *Monocarpus* D. J. Carr

Monocleaceae (Nees) A. B. Frank in Leunis, Syn. Pflanzenk., ed. 2: 1556, 1576. Mar-Dec 1877. Basionym: trib. *Monocleeae* Nees in Gottsche & al., Syn. Hepat.: 507. 5-7 Oct 1846. Type: *Monoclea* Hook.

Monosoleniaceae Inoue in Bull. Natl. Sci. Mus. Tokyo, ser. 2, 9: 117. 15 Jun 1966. Type: *Monosolenium* Griff.

Neotrichocoleaceae Inoue, Ill. Jap. Hepat. 1: 176. 20 Mar 1974. Type: *Neotrichocolea* S. Hatt.

Notothyladaceae (Milde) Müll. Frib. ex Prosk. in Phytomorphology 10: 10. Jun 1960. Basionym: subtrib. *Notothyladinae* Milde in Bot. Zeitung (Berlin) 17: 54. 11 Feb 1859. Type: *Notothylas* Sull. ex A. Gray

Oxymitraceae Müll. Frib. ex Grolle in J. Bryol. 7: 215. 1972. Type: *Oxymitra* Bisch. ex Lindenb.

Pallaviciniaceae Mig., Krypt.-Fl. Deutschl., Moose: 423. Apr 1904. Type: *Pallavicinia* Gray, *nom. cons.*

Pelliaceae H. Klinggr., Höh. Crypt. Preuss.: 13. Apr-Dec 1858. Type: *Pellia* Raddi, *nom. cons.*

Perssoniellaceae R. M. Schust. ex Grolle in J. Bryol. 7: 216. 1972. Type: *Perssoniella* Herzog

Phycolepidoziaceae R. M. Schust. in Bull. Torrey Bot. Club 93: 442. 7 Jan 1967. Type: *Phycolepidozia* R. M. Schust.

Phyllothalliaceae E. A. Hodgs. in Trans. Roy. Soc. New Zealand, Bot. 2: 247. 11 Mai 1964. Type: *Phyllothallia* E. A. Hodgs.

Plagiochilaceae (Jörg.) Müll. Frib. & Herzog in Müller, Leberm. Eur.: 877. Mai 1956. Basionym: trib. *Plagiochileae* Jörg. in Bergens Mus. Skr. 16: 61. 1934. Type: *Plagiochila* (Dumort.) Dumort., *nom. cons.*

Pleuroziaceae (Schiffn.) Müll. Frib., Lebermoose 1: 404. 27 Feb 1909. Basionym: subfam. *Pleurozioideae* Schiffn. in Engler & Prantl, Nat. Pflanzenfam. 1(3): 74, 114. [seors. impr.] Sep 1893. Type: *Pleurozia* Dumort.

Porellaceae Cavers in New Phytol. 9: 292. 1910, *nom. cons.* Type: *Porella* L.

Pseudolepicoleaceae Fulford & J. Taylor in Nova Hedwigia 1: 411. 20 Feb 1960. Type: *Pseudolepicolea* Fulford & J. Taylor

Ptilidiaceae H. Klinggr., Höh. Crypt. Preuss.: 37. Apr-Dec 1858. Type: *Ptilidium* Nees

Radulaceae (Dumort.) Müll. Frib., Lebermoose 1: 404. 27 Feb 1909. Basionym: trib. *Raduleae* Dumort., Recueil Observ. Jungerm.: 13. 1835. Type: *Radula* Dumort., *nom. cons.*

Ricciaceae Rchb., Bot. Damen: 255. 1828. Type: *Riccia* L., *nom. cons. prop.*

Riellaceae Engl., Syllabus, Grosse Ausgabe: 45. Apr 1892. Type: *Riella* Mont.

Sandeothallaceae (R. M. Schust.) R. M. Schust., New Manual Bryol.: 951. 18 Feb 1984. Basionym: subfam. *Sandeothalloideae* R. M. Schust. in Nova Hedwigia 36: 8. 1982. Type: *Sandeothallus* R. M. Schust.

Scapaniaceae Mig., Krypt.-Fl. Deutschl., Moose: 479. Jul 1904. Type: *Scapania* (Dumort.) Dumort., *nom. cons.*

Schistochilaceae H. Buch, Commentat. Biol. 3(1): 9. 1928. Type: *Schistochila* Dumort.

Sphaerocarpaceae (Dumort.) Heeg in Verh. K.K. Zool.-Bot. Ges. Wien 41: 573. 1891. Basionym: trib. *Sphaerocarpeae* Dumort., Jungerm. Europ.: 163. 1874. Type: *Sphaerocarpos* Boehm.

Takakiaceae S. Hatt. & Inoue in J. Hattori Bot. Lab. 19: 137. 7 Mar 1958. Type: *Takakia* S. Hatt. & Inoue

Targioniaceae Dumort., Anal. Fam. Pl.: 68, 70. 1829. Type: *Targionia* L.

Treubiaceae Verd., Man. Bryol.: 427. Sep 1932. Type: *Treubia* K. I. Goebel, *nom. cons.*

Trichocoleaceae Nakai, Chosakuronbun Mokuroku [Ord. Fam. Trib. Nov.]: 201. 20 Jul 1943. Type: *Trichocolea* Dumort., *nom. cons.*

Trichotemnomataceae R. M. Schust. in J. Hattori Bot. Lab. 36: 340. 5 Jan 1973. Type: *Trichotemnoma* R. M. Schust.

Vandiemeniaceae Hewson in J. Hattori Bot. Lab. 52: 163. 29 Jun 1982. Type: *Vandiemenia* Hewson

Vetaformaceae Fulford & J. Taylor in Mem. New York Bot. Gard. 11: 27. 15 Mar 1963. Type: *Vetaforma* Fulford & J. Taylor

Wiesnerellaceae Inoue, Ill. Jap. Hepat. 2: 192. 25 Nov 1976. Type: *Wiesnerella* Schiffn.

FUNGAL FAMILY NAMES IN CURRENT USE

By John C. David

This list is not a catalogue of all validly published fungal family names. For those who need that kind of information, it is available in Hawksworth & David, *Index of fungi supplement – family names* (1989) and in subsequent issues of the *Index of fungi*. It should be well understood that nothing precludes further use of a name that is absent from the following list, but that included names are those for which nomenclatural protection would be appropriate.

For the invidious task of selecting the names to be included it was paramount that no taxonomic prejudice be involved. However, some recent workers have created substantial numbers of new family names, some of which have not been taken up by any other mycologist, perhaps not even used again by their own author. These we have omitted. Conversely we have included some names that are so recent that it may be premature to assert that they are indeed in "current use".

The core of this list comprises those names currently used in *Ainsworth & Bisby's Dictionary of the fungi* (8th edition, in preparation); those accepted in recent issues of *Systema ascomycetum* (Vol. 10); those used by the Mycology Section of the Royal Botanic Gardens, Kew, in the arrangement of their herbarium; and those appearing in relevant review articles for specific groups. Following wide circulation of a first draft of the list, the views of all who commented were duly considered and the necessary additions and amendments were made. Some of the family names listed as rejected in the Berlin (1987) Code, because when families are broadly defined they would otherwise displace a well known family name, can nevertheless be used (as they indeed are) for a narrowly circumscribed family. They are included here, with due mention of their rejected status.

A broad concept of the Fungi has been used for this list, so that certain groups that may be considered to belong elsewhere, such as the Myxomycota, have been covered. However, names of supposed fossil fungi have been excluded as their true nature is in doubt. Also excluded are names based on asexual morphs of fungi with a pleomorphic life cycle (anamorphs), since any such name may only be used to designate a form-taxon of anamorphs and will be displaced by a name typified by a sexual morph (teleomorph) as soon as a family comprising both kinds of morphs is being addressed. Formal classification of anamorphic fungi is in itself unnatural and therefore not used by most mycologists.

The contributions of all those who responded to the circulated draft of this list are acknowledged in the general introduction, but I would like to record my particular thanks to Professor D. L. Hawksworth and Dr. Paul Kirk for assistance with the intricacies of nomenclature. I am also much indebted to Professor W. Greuter for his painstaking attention to the editing of the list.

Acarosporaceae Zahlbr. in Engler & Prantl, Nat. Pflanzenfam. 1(1*): 150. Aug 1906. Type: *Acarospora* A. Massal.

Acaulosporaceae J. B. Morton & Benny in Mycotaxon 37: 73. 23 Apr 1990. Type: *Acaulospora* Gerd. & Trappe

Acrasaceae Poche in Arch. Protistenk. 30: 177. 12 Sep 1913. Type: *Acrasis* Tiegh.

Acrocordiaceae Oksner ex M. E. Barr in Mycotaxon 29: 505. 31 Jul 1987. Type: *Acrocordia* A. Massal.

Acrospermaceae Fuckel in Jahrb. Nassauischen Vereins Naturk. 23-24: 92. Feb-Nov 1870. Type: *Acrospermum* Tode : Fr.

Agaricaceae Chevall., Fl. Gén. Env. Paris 1: 121. 5 Aug 1826. Type: *Agaricus* L. : Fr., *nom. cons.*

Agaricostilbaceae Oberw. & R. Bauer in Sydowia 41: 240. 1 Mar 1989. Type: *Agaricostilbum* J. E. Wright

Agyriaceae Corda, Icon. Fung. 2: 36. Jul 1838. Type: *Agyrium* Fr. : Fr.

Albatrellaceae (Pouzar) Nuss in Hoppea 39: 174. 1980. Basionym: subfam. *Albatrelloideae* Pouzar in Folia Geobot. Phytotax. Bohemoslov. 1: 359. 30 Dec 1966. Type: *Albatrella* Gray

Albuginaceae J. Schröt. in Engler & Prantl, Nat. Pflanzenfam. 1(1): 110. Oct 1893. Type: *Albugo* (Pers.) Roussel ex Gray

Alectoriaceae (Hue) Tomas. in Arch. Bot. (Forlí) 25: 235. 31 Dec 1949. Basionym: trib. *Alectorieae* Hue in Nouv. Arch. Mus. Hist. Nat., ser. 4, 1: 86. 1899. Type: *Alectoria* Ach.

Aleurodiscaceae Jülich in Biblioth. Mycol. 85: 354. 1 Feb 1982. Type: *Aleurodiscus* Rabenh. ex J. Schröt., *nom. cons.*

Amanitaceae R. Heim ex Pouzar in Česká Mykol. 37: 173. 28 Jul 1983. Type: *Amanita* Pers., *nom. cons.*

Amorphothecaceae Parbery in Austral. J. Bot. 17: 345. Aug 1969. Type: *Amorphotheca* Parbery

Amphisphaeriaceae G. Winter in Rabenh. Krypt.-Fl., ed. 2, 1(2): 259. Jul 1885. Type: *Amphisphaeria* Ces. & De Not., *nom. cons.*

Amylariaceae Corner in Beih. Nova Hedwigia 33: 5. Jun 1970. Type: *Amylaria* Corner

Amylocorticaceae Jülich in Biblioth. Mycol. 85: 354. 1 Feb 1982. Type: *Amylocorticum* Pouzar

Antennulariellaceae Woron. in Ann. Mycol. 23: 178. 20 Aug

1925. Type: *Antennulariella* Woron.

Aphelariaceae Corner in Beih. Nova Hedwigia 33: 5. Jun 1970. Type: *Aphelaria* Corner

Aporpiaceae Bondartsev & Bondartseva in Bot. Žurn. (Moscow & Leningrad) 45: 1694. 9 Nov 1960. Type: *Aporpium* Bondartsev & Singer

Arctomiaceae Th. Fr. in Nova Acta Regiae Soc. Sci. Upsal., ser. 3, 3: 287. 1 Apr 1860. Type: *Arctomia* Th. Fr.

Arcyriaceae Rostaf. ex Cooke, Contr. Mycol. Brit.: 69. Jan-Jul 1877. Type: *Arcyria* F. H. Wigg.

Argynnaceae Shearer & J. L. Crane in Trans. Brit. Mycol. Soc. 75: 193. Dec 1980. Type: *Argynna* Morgan

Arthoniaceae Rchb., Deut. Bot. Herb.-Buch: 13, xxvii. Jul 1841. Type: *Arthonia* Ach., *nom. cons.*

Arthopyreniaceae Walt. Watson in New Phytol. 28: 107. 22 Mai 1929. Type: *Arthopyrenia* A. Massal.

Arthrodermataceae Locq. ex Currah in Mycotaxon 24: 36. 11 Dec 1985. Type: *Arthroderma* Curr. ex Berk.

Arthrorhaphidaceae Poelt & Hafellner in Phyton (Horn) 17: 220. 18 Aug 1976. Type: *Arthrorhaphis* Th. Fr.

Ascobolaceae Boud. ex Sacc. in Bot. Centralbl. 18: 219. Mai 1884. Type: *Ascobolus* Pers. : Fr.

Ascocorticiaceae J. Schröt. in Cohn, Krypt.-Fl. Schlesien 3(2): 15. Oct 1893. Type: *Ascocorticium* Bref.

Ascodesmidaceae J. Schröt. in Cohn, Krypt.-Fl. Schlesien 3(2): 31. Oct 1893. Type: *Ascodesmis* Tiegh.

Ascodichaenaceae D. Hawksw. & Sherwood in Mycotaxon 16: 262. 27 Oct 1982. Type: *Ascodichaena* Butin

Ascosphaeraceae L. S. Olive & Spiltoir in Mycologia 47: 242. 29 Apr 1955. Type: *Ascosphaera* L. S. Olive & Spiltoir

Asellariaceae Manier in Ann. Sci. Nat., Bot., ser. 12, 9: 526. Oct 1969. Type: *Asellaria* R. Poiss.

Aspidotheliaceae Räsänen ex J. C. David & D. Hawksw. in Syst. Ascomycetum 10: 13. Jun 1991. Type: *Aspidothelium* Vain.

Asterinaceae Hansf. in Mycol. Pap. 15: 188. 3 Apr 1946. Type: *Asterina* Lév.

Asterostromataceae (Donk) Pouzar in Česká Mykol. 37: 173. 28 Jul 1983. Basionym: subfam. *Asterostromatoideae* Donk in Persoonia 3: 275. 30 Jul 1964. Type: *Asterostroma* Massee

Asterothyriaceae Walt. Watson ex R. Sant. in Symb. Bot. Upsal. 12: 316. 22 Nov 1952. Type: *Asterothyrium* Müll. Arg.

Astraeaceae V. J. Staněk in Novák, Fl. ČSR, Řada B, 1: 803, 818. 1958. Type: *Astraeus* Morgan

Atheliaceae Jülich in Biblioth. Mycol. 85: 355. 1 Feb 1982. Type: *Athelia* Pers.

Atractogloeaceae Oberw. & R. Bauer in Sydowia 41: 245. 1 Mar 1989. Type: *Atractogloea* Oberw. & Bandoni

Auriculariaceae Fr., Epicr. Syst. Mycol.: 530. 1838. Type: *Auricularia* Bull. ex Juss.

Auriscalpiaceae Maas Geest. in Proc. Kon. Ned. Akad. Wetensch., Sect. C, Biol. Sci., 66: 426. 1963. Type: *Auriscalpium* Gray

Bacidiaceae Walt. Watson in New Phytol. 28: 27. 11 Mar 1929. Type: *Bacidia* De Not.

Baeomycetaceae Zenker in Goebel, Pharm. Waarenk. 1: 123. 1827-1829. Type: *Baeomyces* Pers. : Fr.

Balsamiaceae E. Fisch. in Engler & Prantl, Nat. Pflanzenfam. 1(1): 288. Feb 1897. Type: *Balsamia* Vittad.

Bankeraceae Donk in Persoonia 1: 405. 12 Jul 1961. Type: *Bankera* Coker & Beers ex Pouzar

Basidiobolaceae Engl. & Gilg in Engler, Syllabus, ed. 9 & 10: 45. Nov-Dec 1924. Type: *Basidiobolus* Eidam

Batistiaceae (Cif.) Samuels & K. F. Rodrigues in Mycologia 81: 54. 15 Mar 1989. Basionym: subfam. *Batistioideae* Cif. in Atti Ist. Bot. Univ. Pavia, ser. 5, 15: 167. 1958. Type: *Batistia* Cif.

Beenakiaceae Jülich in Biblioth. Mycol. 85: 356. 1 Feb 1982. Type: *Beenakia* D. A. Reid

Bjerkanderaceae Jülich in Biblioth. Mycol. 85: 356. 1 Feb 1982. Type: *Bjerkandera* P. Karst.

Blastocladiaceae H. E. Petersen in Bot. Tidsskr. 29: 357, 395. 5 Dec 1909. Type: *Blastocladia* Reinsch

Bolbitiaceae Singer in Pap. Michigan Acad. Sci. 32: 147. 10 Aug 1950. Type: *Bolbitius* Fr.

Boletaceae Chevall., Fl. Gén. Env. Paris 1: 248. 5 Aug 1826. Type: *Boletus* L. : Fr., *nom. cons.*

Boliniaceae Rick in Brotéria, Sér. Bot. 25: 65. 1 Jul 1931. Type: *Bolinia* (Nitschke) Sacc.

Bondarzewiaceae Kotl. & Pouzar in Česká Mykol. 11: 163. Aug 1957. Type: *Bondarzewia* Singer

Botryobasidiaceae (Parmasto) Jülich in Biblioth. Mycol. 85: 357. 1 Feb 1982. Basionym: subfam. Botryobasidioideae Parmasto, Consp. Syst. Corticiac.: 49. 29 Jul 1968. Type: *Botryobasidium* Donk

Botryosphaeriaceae Theiss. & Syd. in Ann. Mycol. 16: 33. 31 Jul 1918. Type: *Botryosphaeria* Ces. & De Not.

Brachybasidiaceae Gäum., Vergl. Morph. Pilze: 489. 1926. Type: *Brachybasidium* Gäum.

Brigantiaeaceae Hafellner & Bellem. in Nova Hedwigia 35: 246. 1 Jun 1982. Type: *Brigantiaea* Trevis.

Caliciaceae Chevall., Fl. Gén. Env. Paris 1: 314. 5 Aug 1826. Type: *Calicium* Pers. : Fr.

Calosphaeriaceae Munk in Dansk Bot. Ark. 17: 278. 20 Jun 1957. Type: *Calosphaeria* Tul. & C. Tul.

Calycidiaceae Elenkin in Izv. Glavn. Bot. Sada SSSR 28: 267. 1929. Type: *Calycidium* Stirt.

Candelariaceae Hakul. in Ann. Bot. Soc. Zool.-Bot. Fenn. "Vanamo" 27(3): 11. 27 Mar 1954. Type: *Candelaria* A. Massal.

Cantharellaceae J. Schröt. in Cohn, Krypt.-Fl. Schlesien 3(1): 413, 507. 2 Jun 1888. Type: *Cantharellus* Adans. ex Fr. : Fr.

Capnodiaceae (Sacc.) Höhn. in Sitzungsber. Kaiserl. Akad. Wiss., Math.-Naturwiss. Cl., Abt. 1, 118: 1197. 1909. Basionym: subfam. *Capnodioideae* Sacc., Syll. Fung. 1: 73. 13 Jun 1882. Type: *Capnodium* Mont.

Carbomycetaceae Trappe in Trans. Brit. Mycol. Soc. 57: 87. 26 Aug 1971. Type: *Carbomyces* Gilkey

Carcinomycetaceae Oberw. & Bandoni in Nordic J. Bot. 2: 502. 30 Nov 1982. Type: *Carcinomyces* Oberw. & Bandoni

Catabotryaceae Petr. ex M. E. Barr in Mycotaxon 39: 83. 26 Nov 1990. Type: *Catabotrys* Theiss. & P. Syd.

Catenariaceae (Sparrow) Couch in Mycologia 37: 187. 6 Apr 1945. Basionym: subfam. *Catenarioideae* Sparrow, Aquat. Phycomyc.: 317. 1943. Type: *Catenaria* Sorokīn

Catillariaceae Hafellner in Beih. Nova Hedwigia 79: 271. Nov 1984. Type: *Catillaria* A. Massal.

Cavosteliaceae L. S. Olive in Mycologia 56: 885. 2 Feb 1965. Type: *Cavostelium* L. S. Olive

Cejpomycetaceae Jülich in Biblioth. Mycol. 85: 359. 1 Feb 1982. Type: *Cejpomyces* Svrček & Pouzar

Cenangiaceae Bail in Nees & Henry, Syst. Pilze 2: 59. Jan-Feb 1858. Type: *Cenangium* Fr. : Fr.

Cephaloascaceae L. R. Batra in Techn. Bull. U.S.D.A. 1469: 3. Sep 1973. Type: *Cephaloascus* Hanawa

Cephalothecaceae Höhn. in Ann. Mycol. 15: 362. 30 Nov 1917. Type: *Cephalotheca* Fuckel

Ceratiomyxaceae J. Schröt. in Engler & Prantl, Nat. Pflanzenfam. 1(1): 15. Sep 1889. Type: *Ceratiomyxa* J. Schröt.

Ceratobasidiaceae G. W. Martin in Lloydia 11: 114. 1 Dec 1948. Type: *Ceratobasidium* D. P. Rogers

Ceratomycetaceae S. Colla in Fl. Ital. Crypt. 1(16): 134. 30 Jun 1934. Type: *Ceratomyces* Thaxt.

Ceratostomataceae G. Winter in Rabenh. Krypt.-Fl., ed. 2, 1(2): 247. Apr 1885. Type: *Ceratostoma* Fr.

Cerinomycetaceae Jülich in Biblioth. Mycol. 85: 358. 1 Feb 1982. Type: *Cerinomyces* G. W. Martin

Chaconiaceae Cummins & Y. Hirats., Ill. Gen. Rust Fungi, ed. 2: 14. 1983. Type: *Chaconia* Juel

Chadefaudiellaceae Faurel & Schotter ex Benny & Kimbr. in Mycotaxon 12: 46. 16 Dec 1980. Type: *Chadefaudiella* Faurel & Schotter

Chaetocladiaceae J. Schröt. in Engler & Prantl, Nat. Pflanzenfam. 1(1): 131. Dec 1894. Type: *Chaetocladium* Fresen.

Chaetomiaceae G. Winter in Rabenh. Krypt.-Fl., ed. 2, 1(2): 153. Nov 1884. Type: *Chaetomium* Kunze

Chaetothyriaceae Hansf. ex M. E. Barr in Mycologia 71: 943. 20 Nov 1979. Type: *Chaetothyrium* Speg.

Chamonixiaceae Jülich in Biblioth. Mycol. 85: 359. 1 Feb 1982. Type: *Chamonixia* Rolland

Chionosphaeraceae Oberw. & Bandoni in Canad. J. Bot. 60: 1732. Sep 1982. Type: *Chionosphaera* D. E. Cox

Choanephoraceae J. Schröt. in Engler & Prantl, Nat. Pflanzenfam. 1(1): 131. Dec 1894. Type: *Choanephora* Curr.

Chrysotrichaceae Zahlbr. in Engler & Prantl, Nat. Pflanzenfam. 1(1*): 117. Mar 1905. Type: *Chrysothrix* Mont., *nom. cons.*

Chytridiaceae Nowak. in Pamietn. Akad. Umiejetn. w Krakowie, Wydz. Mat.-Przyr., 4: 174, 191. 1878. Type: *Chytridium* A. Braun

Cladochytriaceae J. Schröt. in Engler & Prantl, Nat. Pflanzenfam. 1(1): 80. Aug 1892. Type: *Cladochytrium* Nowak.

Cladoniaceae Zenker in Goebel, Pharm. Waarenk. 1: 124. 1827-1829. Type: *Cladonia* P. Browne, *nom. cons.*

Clastodermataceae Alexop. & T. E. Brooks in Mycologia 63: 926. 25 Aug 1971. Type: *Clastoderma* A. Blytt

Clathraceae Chevall., Fl. Gén. Env. Paris 1: 120. 5 Aug 1826. Type: *Clathrus* L. : Pers.

Clavariaceae Chevall., Fl. Gén. Env. Paris 1: 102. 5 Aug 1826. Type: *Clavaria* L. : Fr., *nom. cons.*

Clavariadelphaceae Corner in Beih. Nova Hedwigia 33: 6. Jun 1970. Type: *Clavariadelphus* Donk

Clavicipitaceae (Lindau) Earle in Contr. U.S. Natl. Herb. 6: 170. 31 Jul 1901. Basionym: subfam. *Clavicipitoideae* Lindau in Engler & Prantl, Nat. Pflanzenfam. 1(1): 348. Apr 1897. Type: *Claviceps* C. Tul.

Clavicoronaceae Corner in Beih. Nova Hedwigia 33: 6. Jun 1970. Type: *Clavicorona* Doty

Clavulinaceae (Donk) Donk in Persoonia 1: 407. 12 Jul 1961. Basionym: trib. *Clavulineae* Donk, Revis. Niederl. Homobasidiomyc.: 16. 7 Jul 1933. Type: *Clavulina* J. Schröt.

Clypeosphaeriaceae G. Winter in Rabenh. Krypt.-Fl., ed. 2, 1(2):

554. Mai 1886. Type: *Clypeosphaeria* Fuckel

Coccocarpiaceae (Mont. ex Müll. Hal.) Henssen in Syst. Ascomycetum 5: 314. Dec 1986. Basionym: trib. *Coccocarpieae* Mont. ex Müll. Hal. in Montagne, Grundr. Flecht.: 26. Apr-Mai 1851. Type: *Coccocarpia* Pers.

Coccodiniaceae Höhn. ex O. E. Erikss. in Opera Bot. 60: 42. Dec 1981. Type: *Coccodinium* A. Massal.

Coccoideaceae Henn. ex Sacc. & D. Sacc. in Saccardo, Syll. Fung. 17: 860. 25 Mai 1905. Type: *Coccoidea* Henn.

Coccotremataceae Henssen ex J. C. David & D. Hawksw. in Syst. Ascomycetum 10: 14. Jun 1991. Type: *Coccotrema* Müll. Arg.

Cochlonemataceae Dudd. in Ainsworth & al., Fungi 4B: 233. 1973. Type: *Cochlonema* Drechsler

Coelomomycetaceae Couch in J. Elisha Mitchell Sci. Soc. 78: 135. Nov 1962. Type: *Coelomomyces* Keilin

Coleosporiaceae Dietel in Engler & Prantl, Nat. Pflanzenfam. 1(1**): 548. Sep 1900. Type: *Coleosporium* Lév.

Collemataceae Zenker in Goebel, Pharm. Waarenk. 1: 124. 1827-1829. Type: *Collema* F. H. Wigg. : Fr., *nom. cons.*

Completoriaceae Humber in Mycotaxon 34: 453. 12 Apr 1989. Type: *Completoria* Lohde

Coniochaetaceae Malloch & Cain in Canad. J. Bot. 49: 878. Jun 1971. Type: *Coniochaeta* (Sacc.) Cooke

Coniocybaceae Rchb., Handb. Nat. Pfl.-Syst.: 132. 1-7 Oct 1837. Type: *Coniocybe* Ach. : Fr.

Coniophoraceae Ulbr. in Lindau, Krypt.-Fl. Anf., ed. 3, 1: 120. Jun-Sep 1928. Type: *Coniophora* DC.

Cookellaceae Höhn. ex Sacc. & Trotter in Saccardo, Syll. Fung. 22: 585. 20 Aug 1913. Type: *Cookella* Sacc.

Coprinaceae Gäum., Vergl. Morph. Pilze: 530. 1926. Type: *Coprinus* Pers. : Fr.

Copromyxaceae L. S. Olive & Stoian., Mycetozoans: 176. 1975. Type: *Copromyxa* Zopf

Coriolaceae (Imazeki) Singer in Publ. Inst. Micol. Univ. Recife 304: 6. 25 Apr 1961. Basionym: subtrib. *Coriolinae* Imazeki in Bull. Tokyo Sci. Mus. 6: 78. 1943. Type: *Coriolus* Quél.

Corticiaceae Herter in Warnstorf & al., Krypt.-Fl. Brandenburg 6: 70. 30 Nov 1910, *nom. cons.* (conserved against *Cyphellaceae* and *Peniophoraceae*). Type: *Corticium* Pers.

Cortinariaceae R. Heim ex Pouzar in Česká Mykol. 37: 174. 28 Jul 1983, *nom. cons.* (conserved against *Crepidotaceae* and *Galeropsidaceae*). Type: *Cortinarius* (Pers.) Gray, *nom. cons.*

Coryneliaceae Sacc. ex Berl. & Voglino in Saccardo, Syll. Fung.

Add. 1-4: 193. 31 Dec 1886. Type: *Corynelia* Ach. ex Fr. : Fr.

Craterellaceae Herter in Warnstorf & al., Krypt.-Fl. Brandenburg 6: 141. 30 Nov 1910. Type: *Craterellus* Pers., *nom. cons.*

Crepidotaceae (S. Imai) Singer in Lilloa 22: 584. 21 Jun 1951 (rejected in favour of *Cortinariaceae*). Basionym: trib. *Crepidoteae* S. Imai in J. Fac. Agric. Hokkaido Univ. 43: 238. Aug 1938. Type: *Crepidotus* (Fr.) Staude

Cribbeaceae Singer in Darwiniana 12: 611. 14 Jun 1963. Type: *Cribbea* A. H. Sm. & D. A. Reid

Cribrariaceae Corda, Icon. Fung. 2: 22. Jul 1838. Type: *Cribraria* Pers.

Crocyniaceae M. Choisy ex Hafellner in Beih. Nova Hedwigia 79: 274. Nov 1984. Type: *Crocynia* (Ach.) A. Massal., *nom. cons.*

Cronartiaceae Dietel in Engler & Prantl, Nat. Pflanzenfam. 1(1**): 548. Sep 1900. Type: *Cronartium* Fr.

Cryptobasidiaceae Malençon ex Donk in Reinwardtia 4: 114. 31 Dec 1956. Type: *Cryptobasidium* Lendn.

Cryptomycocolacaceae Oberw. & R. Bauer in Mycologia 82: 672. 31 Dec 1990. Type: *Cryptomycocolax* Oberw. & R. Bauer

Cucurbitariaceae G. Winter in Rabenh. Krypt.-Fl., ed. 2, 1(2): 308. Jul 1885. Type: *Cucurbitaria* Gray

Cunninghamellaceae Naumov ex R. K. Benj. in Aliso 4: 415. 26 Jun 1959. Type: *Cunninghamella* Matr.

Cyphellaceae Lotsy, Vortr. Bot. Stammesgesch. 1: 695, 696. Jan-Mar 1907 (rejected in favour of *Corticiaceae*). Type: *Cyphella* Fr. : Fr.

Cyttariaceae Speg. in Bol. Acad. Nac. Ci. 11: 253. 1887. Type: *Cyttaria* Berk.

Dacampiaceae Körb., Syst. Lich. Germ.: 322. Jan-Mar 1855. Type: *Dacampia* A. Massal.

Dacrymycetaceae J. Schröt. in Cohn, Krypt.-Fl. Schlesien 3(1): 399. 2 Jun 1888. Type: *Dacrymyces* Nees : Fr.

Dactylosporaceae Bellem. & Hafellner in Cryptog., Mycol., 3: 79. 28 Jun 1982. Type: *Dactylospora* Körb.

Dermeaceae Fr., Summa Veg. Scand.: 345. 1849. Type: *Dermea* Fr.

Dianemataceae T. Macbr., N. Amer. Slime-Moulds: 180. Sep 1899. Type: *Dianema* Rex

Diatrypaceae Nitschke, Pyrenomyc. Germ.: 62. Jan 1867. Type: *Diatrype* Fr.

Dicellomycetaceae Parmasto in Izv. Akad. Nauk. Estonsk. S.S.R., Ser. Biol., 17: 226. 1968. Type: *Dicellomyces* L. S. Olive

Dichostereaceae Jülich in Biblioth. Mycol. 85: 364. 1 Feb 1982. Type: *Dichostereum* Pilát

Dictyonemataceae Tomas. in Arch. Bot. (Forlí) 26: 104. 30 Sep 1950. Type: *Dictyonema* C. Agardh ex Kunth

Dictyosteliaceae Rostaf. ex Cooke, Contr. Mycol. Brit.: 53. Jan-Jul 1877. Type: *Dictyostelium* Bref.

Didymiaceae Rostaf. ex Cooke, Contr. Mycol. Brit.: 29. Jan-Jul 1877. Type: *Didymium* Schrad.

Didymosphaeriaceae Munk in Dansk Bot. Ark. 15: 128. 20 Jun 1953. Type: *Didymosphaeria* Fuckel

Dimargaritaceae R. K. Benj. in Aliso 4: 364. 26 Jun 1959. Type: *Dimargaris* Tiegh.

Dipodascaceae Engl. & Gilg in Engler, Syllabus, ed. 9 & 10: 59. Nov-Dec 1924. Type: *Dipodascus* Lagerh.

Dothideaceae Chevall., Fl. Gén. Env. Paris 1: 446. 5 Aug 1826. Type: *Dothidea* Fr. : Fr.

Dothioraceae Theiss. & Syd. in Ann. Mycol. 15: 444. 30 Apr 1918. Type: *Dothiora* Fr., *nom. cons.*

Echinodontiaceae Donk in Persoonia 1: 405. 12 Jul 1961. Type: *Echinodontium* Ellis & Everh.

Echinosteliaceae Rostaf. ex Cooke, Contr. Mycol. Brit.: 53. Jan-Jul 1877. Type: *Echinostelium* de Bary

Echinosteliopsidaceae L. S. Olive in Bot. Rev. (Lancaster) 36: 83. 26 Jun 1970. Type: *Echinosteliopsis* D. J. Reinh. & L. S. Olive

Ectolechiaceae (Vain.) Zahlbr. in Engler & Prantl, Nat. Pflanzenfam. 1(1*): 122. Mar 1905. Basionym: trib. *Ectolechieae* Vain. in J. Bot. 34: 205. Mai 1896. Type: *Ectolechia* Trevis.

Ectrogellaceae Cejp in Novák, Fl. ČSR, Řada B, 2: 64, 420. 1959. Type: *Ectrogella* Zopf

Eigleraceae Hafellner in Beih. Nova Hedwigia 79: 276. Nov 1984. Type: *Eiglera* Hafellner

Elaeomyxaceae Hagelst. ex M. L. Farr & H. W. Keller in Mycologia 74: 857. 15 Oct 1982. Type: *Elaeomyxa* Hagelst.

Elaphomycetaceae Bail in Nees & Henry, Syst. Pilze 2: 49. Jan-Feb 1858. Type: *Elaphomyces* Nees : Fr.

Elasmomycetaceae Locq. ex Pegler & T. W. K. Young in Trans. Brit. Mycol. Soc. 72: 367. 24 Jul 1979. Type: *Elasmomyces* Cavara

Elsinoaceae Höhn. ex Sacc. & Trotter in Saccardo, Syll. Fung. 22: 584. 20 Aug 1913. Type: *Elsinoë* Racib.

Endemosarcaceae L. S. Olive & Erdos in Mycologia 63: 877. 25 Aug 1971. Type: *Endemosarca* L. S. Olive & Erdos

Endogonaceae Sacc., Syll. Fung. 8: 905. 20 Dec 1889. Type: *Endogone* Link : Fr.

Endomycetaceae J. Schröt. in Cohn, Krypt.-Fl. Schlesien 3(2): 208. Dec 1893. Type: *Endomyces* Reess

Englerulaceae Henn. in Hedwigia 43: 353, 354. 3 Sep 1904. Type: *Englerula* Henn.

Enteridiaceae M. L. Farr in Mycologia 74: 339. 2 Apr 1982. Type: *Enteridium* Ehrenb.

Entolomataceae Kotl. & Pouzar in Česká Mykol. 26: 218. 15 Oct 1972. Type: *Entoloma* (Fr.) P. Kumm.

Entomophthoraceae G. Winter in Rabenh. Krypt.-Fl., ed. 2, 1(1): 71. Dec 1880. Type: *Entomophthora* Fresen.

Eoterfeziaceae G. F. Atk. in Bot. Gaz. (Crawfordsville) 34: 40. 16 Jul 1902. Type: *Eoterfezia* G. F. Atk.

Epigloeaceae Zahlbr. in Engler & Prantl, Nat. Pflanzenfam. 1(1*): 53. Oct 1903. Type: *Epigloea* Zukal

Epithelaceae Jülich in Biblioth. Mycol. 85: 366. 1 Feb 1982. Type: *Epithele* (Pat.) Pat.

Eremomycetaceae Malloch & Cain in Canad. J. Bot. 49: 847. Jun 1971. Type: *Eremomyces* Malloch & Cain

Erysiphaceae G. Winter in Rabenh. Krypt.-Fl., ed. 2, 1(2): 22. Mar 1884. Type: *Erysiphe* R. Hedw. ex DC. : Fr.

Euantennariaceae S. Hughes & Corlett in New Zealand J. Bot. 10: 238. 28 Aug 1972. Type: *Euantennaria* Speg.

Euceratomycetaceae I. I. Tav. in Mycotaxon 11: 488. 13 Aug 1980. Type: *Euceratomyces* Thaxt.

Exobasidiaceae J. Schröt. in Cohn, Krypt.-Fl. Schlesien 3(1): 413. 2 Jun 1888. Type: *Exobasidium* Woronin

Fenestellaceae M. E. Barr in Mycologia 71: 952. 20 Nov 1979. Type: *Fenestella* Tul. & C. Tul.

Filobasidiaceae L. S. Olive in J. Elisha Mitchell Sci. Soc. 84: 261. Jul-Aug 1968. Type: *Filobasidium* L. S. Olive

Fistulinaceae Lotsy, Vortr. Bot. Stammesgesch. 1: 695, 704. Jan-Mar 1907. Type: *Fistulina* Bull. : Fr.

Fomitopsidaceae Jülich in Biblioth. Mycol. 85: 367. 1 Feb 1982. Type: *Fomitopsis* P. Karst.

Fonticulaceae Worley & al. in Mycologia 71: 747. 6 Aug 1979. Type: *Fonticula* Worley & al.

Fuscideaceae Hafellner in Beih. Nova Hedwigia 79: 278. Nov 1984. Type: *Fuscidea* V. Wirth & Vězda

Galeropsidaceae Singer in Bol. Soc. Argent. Bot. 10: 61. 15 Aug 1962 (rejected in favour of *Cortinariaceae*). Type: *Galeropsis* Velen.

Ganodermataceae (Donk) Donk in Bull. Jard. Bot. Buitenzorg, ser. 3, 17: 474. 1948. Basionym: subfam. *Ganodermatoideae* Donk, Revis. Niederl. Homobasidiomyc.: 229. 7 Jul 1933. Type: *Ganoderma* P. Karst.

Gasterellaceae Zeller in Mycologia 40: 639. 23 Dec 1948. Type: *Gasterella* Zeller & L. B. Walker

Gautieriaceae Zeller in Mycologia 40: 666. 23 Dec 1948. Type: *Gautieria* Vittad., *nom. cons.*

Geastraceae Corda, Anleit. Stud. Mykol.: 104. Aug-Dec 1842. Type: *Geastrum* Pers. : Pers.

Gelopellaceae Zeller in Mycologia 31: 20. 1 Feb 1939. Type: *Gelopellis* Zeller

Geneaceae Trappe in Mycotaxon 9: 330. 12 Jun 1979. Type: *Genea* Vittad.

Geoglossaceae Corda, Icon. Fung. 2: 35. Jul 1838. Type: *Geoglossum* Pers. : Fr.

Geosiphonaceae Engl. & Gilg in Engler, Syllabus, ed. 9 & 10: 24. Nov-Dec 1924. Type: *Geosiphon* Wettst.

Gigasporaceae J. B. Morton & Benny in Mycotaxon 37: 483. 23 Apr 1990. Type: *Gigaspora* Gerd. & Trappe

Gilbertellaceae Benny in Mycologia 83: 151. 26 Apr 1991. Type: *Gilbertella* Hesselt.

Glaziellaceae J. L. Gibson in Mycologia 78: 953. 30 Dec 1986. Type: *Glaziella* Berk.

Gloeocystidiellaceae (Parmasto) Jülich in Biblioth. Mycol. 85: 368. 1 Feb 1982. Basionym: subfam. *Gloeocystidielloideae* Parmasto, Consp. Syst. Corticiac.: 134. 29 Jul 1968. Type: *Gloeocystidiellum* Donk

Glomaceae Piroz. & Dalpé in Symbiosis 7: 19. 1989. Type: *Glomus* Tul. & C. Tul.

Gnomoniaceae G. Winter in Rabenh. Krypt.-Fl., ed. 2, 1(2): 570. Mai 1886. Type: *Gnomonia* Ces. & De Not., *nom. cons. prop.*

Gomphaceae Donk in Persoonia 1: 406. 12 Jul 1961. Type: *Gomphus* Pers.

Gomphidiaceae Maire ex Singer in Sydowia 15: 82. 20 Sep 1962. Type: *Gomphidius* Fr.

Gomphillaceae Walt. Watson ex Hafellner in Beih. Nova Hedwigia 79: 280. Nov 1984. Type: *Gomphillus* Nyl.

Grammothelaceae Jülich in Biblioth. Mycol. 85: 369. 1 Feb 1982. Type: *Grammothele* Berk. & M. A. Curtis

Graphidaceae Dumort., Comment. Bot.: 69, 78. Nov-Dec 1822. Type: *Graphis* Adans. : Fr.

Graphiolaceae Clem. & Shear, Gen. Fung., ed. 2: 156. Jun-Jul 1931. Type: *Graphiola* Poit.

Guttulinaceae Zopf ex Berl. in Saccardo, Syll. Fung. 7(1): 451. 15 Mar 1888. Type: *Guttulina* Cienk.

Guttulinopsidaceae L. S. Olive in Bot. Rev. (Lancaster) 36: 75. 26 Jun 1970. Type: *Guttulinopsis* Olive

Gyalectaceae Stizenb. in Ber. Thätigk. St. Gallischen Naturwiss. Ges. 1861-1862: 158. 1862. Type: *Gyalecta* Ach.

Gymnoascaceae Baran. in Bot. Zeitung (Berlin) 30: 158. 8 Mar 1872. Type: *Gymnoascus* Baran.

Gypsoplacaceae Timdal in Biblioth. Lichenol. 38: 423. Jan-Sep 1990. Type: *Gypsoplaca* Timdal

Gyrodontaceae (Singer) Heinem. in Bull. Jard. Bot. Etat 21: 228. Dec 1951. Basionym: trib. *Gyrodonteae* Singer in Farlowia 2: 230. 17 Sep 1945. Type: *Gyrodon* Opat., *nom. cons.*

Haddowiaceae Jülich in Biblioth. Mycol. 85: 370. 1 Feb 1982. Type: *Haddowia* Steyaert

Haematommataceae Hafellner in Beih. Nova Hedwigia 79: 281. Nov 1984. Type: *Haematomma* A. Massal.

Haliphthoraceae Vishniac in Mycologia 50: 75. 4 Apr 1958. Type: *Haliphthoros* Vishniac

Halosphaeriaceae E. Müll. & Arx ex Kohlm. in Canad. J. Bot. 50: 1951. 10 Jan 1972. Type: *Halosphaeria* Linder

Helicocephalaceae Boedijn in Sydowia 12: 355. 25 Mar 1959. Type: *Helicocephalum* Thaxt.

Helvellaceae Fr., Syst. Mycol. 2: 1. 1822. Type: *Helvella* L. : Fr.

Hemiphacidiaceae Korf in Mycologia 54: 26. 5 Oct 1962. Type: *Hemiphacidium* Korf

Heppiaceae Zahlbr. in Engler & Prantl, Nat. Pflanzenfam. 1(1*): 176. Aug 1906. Type: *Heppia* Nägeli ex A. Massal.

Hericiaceae Donk in Persoonia 3: 269. 30 Jul 1964. Type: *Hericium* Pers.

Herpomycetaceae (Thaxt.) I. I. Tav. in Mycotaxon 13: 469. 31 Oct 1981. Basionym: trib. *Herpomyceteae* Thaxt. in Mem. Amer. Acad. Arts 13: 237. 1908. Type: *Herpomyces* Thaxt.

Herpotrichiellaceae Munk in Dansk Bot. Ark. 15: 131. 20 Jun 1953. Type: *Herpotrichiella* Petr.

Heterodeaceae Filson in Lichenologist 10: 15. Apr 1978. Type: *Heterodea* Nyl.

Heterogastridiaceae Oberw. & R. Bauer in Mycologia 82: 57. 27 Feb 1990. Type: *Heterogastridium* Oberw. & R. Bauer

Hoehnelomycetaceae Jülich in Biblioth. Mycol. 85: 371. 1 Feb 1982. Type: *Hoehnelomyces* Weese

Hyaloriaceae Lindau in Engler & Prantl, Nat. Pflanzenfam. 1(1**): 95. Dec 1897. Type: *Hyaloria* Møller

Hyaloscyphaceae Nannf. in Nova Acta Regiae Soc. Sci. Upsal., ser. 4, 8: 258. Sep 1932. Type: *Hyaloscypha* Boud.

Hydnaceae Chevall., Fl. Gén. Env. Paris 1: 270. 5 Aug 1826. Type: *Hydnum* L. : Fr., *nom. cons.*

Hydnangiaceae Gäum. & C. W. Dodge, Compar. Morph. Fungi: 485. 1928 (rejected in favour of *Tricholomataceae*). Type: *Hydnangium* Wallr.

Hygrophoraceae Lotsy, Vortr. Bot. Stammesgesch. 1: 706. Jan-Mar 1907. Type: *Hygrophorus* Fr.

Hygrophoropsidaceae Kühner in Bull. Mens. Soc. Linn. Lyon, Soc. Bot. 49: 414. Sep 1980 (rejected in favour of *Tricholomataceae*).

Type: *Hygrophoropsis* (J. Schröt.) Maire ex Martin-Sans

Hymenangiaceae Corda, Icon. Fung. 5: 28. Jun 1842. Type: *Hymenangium* Klotzsch

Hymeneliaceae Körb., Syst. Lich. Germ.: 327. Jan-Mar 1855. Type: *Hymenelia* Kremp.

Hymenochaetaceae Donk in Bull. Jard. Bot. Buitenzorg, ser. 3, 17: 474. 1948. Type: *Hymenochaete* Lév., *nom. cons.*

Hymenogasteraceae Vittad., Monogr. Tuberac.: 11. 1831. Type: *Hymenogaster* Vittad.

Hyphochytriaceae A. Fisch. in Rabenh. Krypt.-Fl., ed. 2, 1(4): 131. Jan 1892. Type: *Hyphochytrium* Zopf

Hyphodermataceae Jülich in Biblioth. Mycol. 85: 373. 1 Feb 1982. Type: *Hyphoderma* Wallr.

Hypocreaceae De Not. in Giorn. Bot. Ital. 2: 48. 1844. Type: *Hypocrea* Fr.

Hypodermataceae Rehm in Rabenh. Krypt.-Fl., ed. 2, 1(3): 28. Aug 1887. Type: *Hypoderma* De Not., *nom. cons.*

Hypomycetaceae (Lindau) Earle in Contr. U.S. Natl. Herb. 6: 168. 31 Jul 1901. Basionym: subfam. *Hypomycetoideae* Lindau in Engler & Prantl, Nat. Pflanzenfam. 1(1): 346. Apr 1897. Type: *Hypomyces* (Fr.) Tul.

Hyponectriaceae Petr. in Ann. Mycol. 21: 305. 28 Jul 1923. Type: *Hyponectria* Sacc.

Hysterangiaceae E. Fisch. in Engler & Prantl, Nat. Pflanzenfam. 1(1**): 304. Jun 1899. Type: *Hysterangium* Vittad.

Hysteriaceae Chevall., Fl. Gén. Env. Paris 1: 432. 5 Aug 1826. Type: *Hysterium* Pers. : Fr.

Karstenellaceae Harmaja in Karstenia 14: 110. 19 Apr 1974. Type: *Karstenella* Harmaja

Kathistaceae Malloch & M. Blackw. in Canad. J. Bot. 68: 1719. 19 Sep 1990. Type: *Kathistes* Malloch & M. Blackw.

Kickxellaceae Linder in Farlowia 1: 56. 1 Feb 1943. Type: *Kickxella* Coem.

Koralionastaceae Kohlm. & Volkm.-Kohlm. in Mycologia 79: 764. 29 Oct 1987. Type: *Koralionastes* Kohlm. & Volkm.-Kohlm.

Kriegeriellaceae M. E. Barr in Mycotaxon 29: 502. 31 Jul 1987. Type: *Kriegeriella* Höhn.

Krieglsteineraceae Pouzar in Beitr. Kenntn. Pilze Mitteleuropas 3: 404. 8 Sep 1987. Type: *Krieglsteinera* Pouzar

Laboulbeniaceae G. Winter in Rabenh. Krypt.-Fl., ed. 2, 1(2): 918. Mar 1887. Type: *Laboulbenia* Mont. & C. P. Robin

Labyrinthulaceae Cienk. in Arch. Mikroskop. Anat. 3: 274. 1867. Type: *Labyrinthula* Cienk.

Lachnocladiaceae D. A. Reid in Beih. Nova Hedwigia 18: 45. 15 Mai 1965. Type: *Lachnocladium* Lév., *nom. cons.*

Lagenidiaceae J. Schröt. in Engler & Prantl, Nat. Pflanzenfam. 1(1): 89. Oct 1893. Type: *Lagenidium* Schenk

Lahmiaceae O. E. Erikss. in Mycotaxon 27: 357. 3 Dec 1986. Type: *Lahmia* Körb.

Lasiosphaeriaceae Nannf. in Nova Acta Regiae Soc. Sci. Upsal., ser. 4, 8: 50. Sep 1932. Type: *Lasiosphaeria* Ces. & De Not.

Lecanoraceae Körb., Syst. Lich. Germ.: 104. Jan-Mar 1855. Type: *Lecanora* Ach.

Lecideaceae Chevall., Fl. Gén. Env. Paris 1: 549. 5 Aug 1826. Type: *Lecidea* Ach.

Legeriomycetaceae Pouzar in Folia Geobot. Phytotax. 7: 319. 30 Nov 1972. Type: *Legeriomyces* Pouzar

Lentariaceae Jülich in Biblioth. Mycol. 85: 375. 1 Feb 1982. Type: *Lentaria* Corner

Lentinaceae Jülich in Biblioth. Mycol. 85: 376. 1 Feb 1982 (rejected in favour of *Tricholomataceae*). Type: *Lentinus* Fr. : Fr.

Lentinellaceae Kotl. & Pouzar in Česká Mykol. 26: 218. 15 Oct 1972. Type: *Lentinellus* P. Karst.

Leotiaceae Corda, Icon. Fung. 5: 37. Jun 1842. Type: *Leotia* Pers. : Fr.

Leprariaceae Rchb., Deut. Bot. Herb.-Buch: xxvii. Jul 1841. Type: *Lepraria* Ach., *nom. cons.*

Leptolegniellaceae M. W. Dick in Trans. Brit. Mycol. Soc. 57: 421. 24 Dec 1971. Type: *Leptolegniella* Huneycutt

Leptomitaceae Kütz., Phycol. General.: 150. 14-16 Sep 1843. Type: *Leptomitus* C. Agardh

Leptopeltidaceae Höhn. ex Trotter in Saccardo, Syll. Fung. 24: 1255. 25 Apr 1928. Type: *Leptopeltis* Höhn.

Leptosphaeriaceae M. E. Barr in Mycotaxon 29: 503. 31 Jul 1987. Type: *Leptosphaeria* Ces. & De Not., *nom. cons.*

Letrouitiaceae Hafellner & Bellem. in Nova Hedwigia 35: 382. 1 Jun 1982. Type: *Letrouitia* Hafellner & Bellem.

Leucogasteraceae Moreau ex Fogel in Canad. J. Bot. 57: 1723. 1 Aug 1979. Type: *Leucogaster* R. Hesse

Liceaceae Chevall., Fl. Gén. Env. Paris 1: 343. 5 Aug 1826. Type: *Licea* Schrad.

Lichenotheliaceae Henssen in Syst. Ascomycetum 5: 137. Sep 1986. Type: *Lichenothelia* D. Hawksw.

Lichinaceae Nyl. in Mém. Soc. Sci. Nat. Cherbourg 2: 8. Jan-Mar 1854. Type: *Lichina* C. Agardh, *nom. cons.*

Lindtneriaceae Jülich in Biblioth. Mycol. 85: 377. 1 Feb 1982. Type: *Lindtneria* Pilát

Lipomycetaceae E. K. Novák & Zsolt in Acta Bot. Acad. Sci. Hung. 7: 97. 3 Mai 1961. Type: *Lipomyces* Lodder & Kreger

Listerellaceae E. Jahn in Ber. Deutsch. Bot. Ges. 24: 538, 541. 24 Jan 1906. Type: *Listerella* E. Jahn

Lobariaceae Chevall., Fl. Gén. Env. Paris 1: 609. 5 Aug 1826. Type: *Lobaria* (Schreb.) Hoffm.

Lophiaceae H. Zogg ex Arx & E. Müll. in Stud. Mycol. 9: 60. 10 Mar 1975. Type: *Lophium* Fr. : Fr.

Lophiostomataceae Sacc., Syll. Fung. 2: 672. 13 Jun 1883. Type: *Lophiostoma* Ces. & De Not., *nom. cons.*

Loramycetaceae Dennis ex Digby & Goos in Mycologia 79: 829. 19 Jan 1988. Type: *Loramyces* W. Weston

Lycoperdaceae Chevall., Fl. Gén. Env. Paris 1: 348. 5 Aug 1826. Type: *Lycoperdon* Pers. : Pers.

Massariaceae Nitschke in Verh. Naturhist. Vereines Preuss. Rheinl. Westphalens 26: 73. 1869. Type: *Massaria* De Not.

Mastodiaceae Zahlbr. in Engler & Prantl, Nat. Pflanzenfam. 1(1*): 240. Jan 1908. Type: *Mastodia* Hook. f. & Harv.

Medeolariaceae Korf in Mycotaxon 15: 231. 15 Jul 1982. Type: *Medeolaria* Thaxt.

Megachytriaceae Sparrow, Aquat. Phycomyc.: 378. 1943. Type: *Megachytrium* Sparrow

Megalosporaceae Vězda ex Hafellner & Bellem. in Nova Hedwigia 35: 216. 1 Jun 1982. Type: *Megalospora* Meyen

Melampsoraceae Dietel in Engler & Prantl, Nat. Pflanzenfam. 1(1**): 38. Oct 1897. Type: *Melampsora* Castagne

Melanconidaceae G. Winter in Rabenh. Krypt.-Fl., ed. 2, 1(2): 764. Oct 1886. Type: *Melanconis* Tul. & C. Tul.

Melanogasteraceae E. Fisch. in Engler & Prantl, Nat. Pflanzenfam., ed. 2, 7a: 9. Jan 1933. Type: *Melanogaster* Corda, *nom. cons.*

Melanommataceae G. Winter in Rabenh. Krypt.-Fl., ed. 2, 1(2): 220. Apr 1885. Type: *Melanomma* Nitschke ex Fuckel

Melaspileaceae Walt. Watson in New Phytol. 28: 94. 22 Mai 1929. Type: *Melaspilea* Nyl.

Meliolaceae G. W. Martin ex Hansf. in Mycol. Pap. 15: 23. 3 Apr 1946. Type: *Meliola* Fr. : Fr.

Melogrammataceae G. Winter in Rabenh. Krypt.-Fl., ed. 2, 1(2): 797. Oct 1886. Type: *Melogramma* Fr.

Meristacraceae Humber in Mycotaxon 34: 456. 12 Apr 1989. Type: *Meristacrum* Drechsler

Meruliaceae Rea, Brit. Basidiomyc.: 10, 620. 1 Aug 1922. Type: *Merulius* Fr. : Fr.

Mesnieraceae Arx & E. Müll. in Stud. Mycol. 9: 94. 10 Mar 1975. Type: *Mesniera* Sacc. & Syd.

Metacapnodiaceae S. Hughes & Corlett in New Zealand J. Bot. 10: 239. 28 Aug 1972. Type: *Metacapnodium* Speg.

Metschnikowiaceae T. Kamieński in Trudy Imp. S.-Peterburgsk. Obšč. Estestvoisp., Vyp. 3, Otd. Bot. 30: 345. 1899. Type: *Metschnikowia* T. Kamieński

Micareaceae Vězda ex Hafellner in Beih. Nova Hedwigia 79: 306. Nov 1984. Type: *Micarea* Fr.

Microascaceae Luttr. ex Malloch in Mycologia 62: 734. 11 Sep 1970. Type: *Microascus* Zukal

Microcaliciaceae Tibell in Beih. Nova Hedwigia 79: 701. Nov 1984. Type: *Microcalicium* Vain.

Micromycopsidaceae Subram. in Curr. Sci. 43: 723. 20 Nov 1974. Type: *Micromycopsis* Scherff.

Micropeltidaceae Clem. & Shear, Gen. Fung., ed. 2: 100. Jun-Jul 1931. Type: *Micropeltis* Mont.

Microtheliopsidaceae O. E. Erikss. in Opera Bot. 60: 97. Dec 1981. Type: *Microtheliopsis* Müll. Arg.

Microthyriaceae Sacc., Syll. Fung. 2: 658. 13 Jun 1883. Type: *Microthyrium* Desm.

Mikronegeriaceae Cummins & Y. Hirats., Ill. Gen. Rust Fungi, ed. 2: 13. 1983. Type: *Mikronegeria* Dietel

Miltideaceae Hafellner in Beih. Nova Hedwigia 79: 308. Nov 1984. Type: *Miltidea* Stirt.

Mixiaceae C. L. Kramer in Stud. Mycol. 30: 159. 31 Dec 1987. Type: *Mixia* C. L. Kramer

Monascaceae J. Schröt. in Engler & Prantl, Nat. Pflanzenfam. 1(1): 148. Dec 1894. Type: *Monascus* Tiegh.

Monoblastiaceae Walt. Watson in New Phytol. 28: 106. 22 Mai 1929. Type: *Monoblastia* Riddle

Monoblepharidaceae A. Fisch. in Rabenh. Krypt.-Fl., ed. 2, 1(4): 378. Mai 1892. Type: *Monoblepharis* Cornu

Montagneaceae Singer in Rev. Mycol. (Paris) 40: 61. Jun 1976. Type: *Montagnea* Fr.

Morchellaceae Rchb., Pflanzenreich: 2. 1834. Type: *Morchella* Pers. : Fr.

Moriolaceae Zahlbr. in Engler & Prantl, Nat. Pflanzenfam. 1(1*): 52. Oct 1903. Type: *Moriola* Norman

Mortierellaceae A. Fisch. in Rbenh. Krypt.-Fl., ed. 2, 1(4): 268. Apr 1892. Type: *Mortierella* Coem.

Mucoraceae Dumort., Comment. Bot.: 69, 81. Nov-Dec 1822. Type: *Mucor* Fresen., *nom. cons.*

Mycoblastaceae Hafellner in Beih. Nova Hedwigia 79: 310. Nov 1984. Type: *Mycoblastus* Norman

Mycocaliciaceae A. F. W. Schmidt in Mitt. Staatsinst. Allg. Bot. Hamburg 13: 127. Dec 1970. Type: *Mycocalicium* Vain.

Mycoporaceae Zahlbr. in Engler & Prantl, Nat. Pflanzenfam. 1(1*): 77. Oct 1903. Type: *Mycoporum* Flot. ex Nyl., *nom. cons.*

Mycosphaerellaceae Lindau in Engler & Prantl, Nat. Pflanzenfam. 1(1): 421. Jun 1897. Type: *Mycosphaerella* Johanson

Mycotyphaceae Benny & R. K. Benj. in Mycotaxon 22: 122. 8 Feb 1985. Type: *Mycotypha* Fenner

Myriangiaceae Nyl. in Mém. Soc. Sci. Nat. Cherbourg 2: 9. Jan-Mar

1854. Type: *Myriangium* Mont. & Berk.

Mytilinidiaceae Kirschst. in Verh. Bot. Vereins Prov. Brandenburg 66: 28. 1924. Type: *Mytilinidion* Duby

Myxotrichaceae Locq. ex Currah in Mycotaxon 24: 103. 11 Dec 1985. Type: *Myxotrichum* Kunze

Neocallimastigaceae I. B. Heath in Canad. J. Bot. 61: 304. 8 Mar 1983. Type: *Neocallimastix* Vavra & Joyon ex I. B. Heath

Neolectaceae Redhead in Canad. J. Bot. 55: 305. 1 Feb 1977. Type: *Neolecta* Speg.

Nephromataceae Wetmore ex J. C. David & D. Hawksw. in Syst. Ascomycetum 10: 15. Jun 1991. Type: *Nephroma* Ach.

Nidulariaceae Dumort., Comment. Bot.: 69, 81. Nov-Dec 1822. Type: *Nidularia* Fr., *nom. cons.*

Niessliaceae Kirschst. in Ann. Mycol. 37: 89. 30 Apr 1939. Type: *Niesslia* Auersw.

Nitschkiaceae (Fitzp.) Nannf. in Nova Acta Regiae Soc. Sci. Upsal., ser. 4, 8: 56. Sep 1932. Basionym: subfam. *Nitschkioideae* Fitzp. in Mycologia 15: 26. 25 Jan 1923. Type: *Nitschkia* G. H. Otth ex P. Karst.

Octavianinaceae Locq. ex Pegler & T. W. K. Young in Trans. Brit. Mycol. Soc. 72: 379. 24 Jul 1979. Type: *Octavianina* Kuntze

Odontotremataceae D. Hawksw. & Sherwood in Mycotaxon 16: 263. 27 Oct 1982. Type: *Odontotrema* Nyl.

Oedogoniomycetaceae D. J. S. Barr in Margulis & al., Handb. Protoctista: 463. 1991. Type: *Oedogoniomyces* Tak. Kobay. & M. Ôkubo

Olpidiaceae J. Schröt. in Cohn, Krypt.-Fl. Schlesien 3(1): 180. 16 Aug 1886. Type: *Olpidium* (A. Braun) J. Schröt.

Olpidiopsidaceae Sparrow ex Cejp in Novák, Fl. ČSR, Řada B, 2: 349, 455. 1959. Type: *Olpidiopsis* Cornu

Onygenaceae Berk., Intr. Crypt. Bot.: 272. Mar-Jun 1857. Type: *Onygena* Pers. : Fr.

Opegraphaceae Stizenb. in Ber. Thätigk. St. Gallischen Naturwiss. Ges. 1861-1862: 153. 1862. Type: *Opegrapha* Ach. : Fr., *nom. cons.*

Ophioparmaceae R. W. Rogers & Hafellner in Lichenologist 20: 172. April 1988. Type: *Ophioparma* Norman

Ophiostomataceae Nannf. in Nova Acta Regiae Soc. Sci. Upsal., ser. 4, 8: 30. Sep 1932. Type: *Ophiostoma* Syd. & P. Syd.

Orbiliaceae Nannf. in Nova Acta Regiae Soc. Sci. Upsal., ser. 4, 8: 250. Sep 1932. Type: *Orbilia* Fr.

Otideaceae Eckblad in Nytt Mag. Bot. 15: 82. 15 Jun 1968. Type: *Otidea* (Pers.) Bonord.

Pachnocybaceae Oberw. & R. Bauer in Sydowia 41: 244. 1 Mar 1989. Type: *Pachnocybe* Berk.

Palavasciaceae Manier in Ann. Sci. Nat., Bot., ser. 12, 9: 525. Oct 1969. Type: *Palavascia* Tuzet & Manier ex Lichtw.

Pannariaceae Tuck., Gen. Lich.: xii. Jun-Dec 1872. Type: *Pannaria* Delise ex Bory

Parataeniellaceae Manier & Lichtw. in Ann. Sci. Nat., Bot., ser. 12, 9: 525. Oct 1969. Type: *Parataeniella* R. Poiss.

Parmeliaceae Zenker in Goebel, Pharm. Waarenk. 1: 124. 1827-1829. Type: *Parmelia* Ach. : Fr., *nom. cons.*

Parmulariaceae E. Müll. & Arx ex M. E. Barr in Mycologia 71: 944. 20 Nov 1979. Type: *Parmularia* Lév.

Parodiellaceae Theiss. & P. Syd. ex M. E. Barr in Mycotaxon 29: 503. 31 Jul 1987. Type: *Parodiella* Speg.

Parodiopsidaceae Toro in J. Agric. Univ. Puerto Rico 36: 66. Jan 1952. Type: *Parodiopsis* Maubl.

Patellariaceae Corda, Icon. Fung. 2: 37. Jul 1838. Type: *Patellaria* Fr. : Fr.

Paxillaceae Lotsy, Vortr. Bot. Stammesgesch. 1: 706. Jan-Mar 1907. Type: *Paxillus* Fr.

Peltigeraceae Dumort., Comment. Bot.: 68, 78. Nov-Dec 1822. Type: *Peltigera* Willd., *nom. cons.*

Peltulaceae Büdel in Syst. Ascomycetum 5: 149. Sep 1986. Type: *Peltula* Nyl.

Peniophoraceae Lotsy, Vortr. Bot. Stammesgesch. 1: 687, 689. Jan-Mar 1907 (rejected in favour of *Corticiaceae*). Type: *Peniophora* Cooke

Peronophythoraceae W. H. Ko & al. in Mycologia 70: 383. 25 Mai 1978. Type: *Peronophythora* C. C. Chen ex W. H. Ko & al.

Peronosporaceae Warm., Haandb. Syst. Bot., ed. 2: 54. Jan-Sep 1884. Type: *Peronospora* Corda

Pertusariaceae Körb., Syst. Lich. Germ.: 377. Jan-Mar 1855. Type: *Pertusaria* DC., *nom. cons.*

Pezizaceae Dumort., Anal. Fam. Pl.: 72. 1829. Type: *Peziza* Fr. : Fr.

Phacidiaceae Fr., Summa Veg. Scand.: 367. 1849. Type: *Phacidium* Fr. : Fr., *nom. cons.*

Phaeosphaeriaceae M. E. Barr in Mycologia 71: 948. 20 Nov 1979. Type: *Phaeosphaeria* I. Miyake

Phaeotrichaceae Cain in Canad. J. Bot. 34: 676. Jul 1956. Type: *Phaeotrichum* Cain & M. E. Barr

Phakopsoraceae (Arthur) Cummins & Y. Hirats., Ill. Gen. Rust Fungi, ed. 2: 13. 1983. Basionym: trib. *Phakopsoreae* Arthur, Man. Rusts: 1. 1934. Type: *Phakopsora* Dietel

Phallaceae Corda, Icon. Fung. 5: 29. Jun 1842. Type: *Phallus* L. : Pers.

Phillipsiellaceae Höhn. in Sitzungsber. Kaiserl. Akad. Wiss., Math.-Naturwiss. Cl., Abt. 1, 118: 361. 1909. Type: *Phillipsiella* Cooke

Phleogenaceae Weese in Ber. Deutsch. Bot. Ges. 37: 518. 25 Feb 1920. Type: *Phleogena* Link

Phlyctidaceae Poelt ex J. C. David & D. Hawksw. in Syst. Ascomycetum 10: 15. Jun 1991. Type: *Phlyctis* (Wallr.) Flot., *nom. cons.*

Phragmidiaceae Corda, Icon. Fung. 1: 6. Aug 1837. Type: *Phragmidium* Link

Phragmopelthecaceae L. Xavier, Phragmopeltecac.: 42. 1976. Type: *Phragmopeltheca* L. Xavier

Phyllachoraceae Theiss. & Syd. in Ann. Mycol. 13: 168. 31 Aug 1915. Type: *Phyllachora* Nitschke ex Fuckel, *nom. cons.*

Phyllobatheliaceae Bitter & F. Schill. in Hedwigia 67: 272, 291. Dec 1927. Type: *Phyllobathelium* (Müll. Arg.) Müll. Arg.

Physalacriaceae Corner in Beih. Nova Hedwigia 33: 10. Jun 1970 (rejected in favour of *Tricholomataceae*). Type: *Physalacria* Peck

Physaraceae Chevall., Fl. Gén. Env. Paris 1: 332. 5 Aug 1826. Type: *Physarum* Pers.

Physciaceae Zahlbr. in Engler, Syllabus, ed. 2: 46. Mai 1898. Type: *Physcia* (Schreb.) Michx.

Physodermataceae Sparrow in Mycologia 44: 768. 3 Jan 1953. Type: *Physoderma* Wallr.

Piedraiaceae Viégas ex Cif. & al. in Publ. Inst. Micol. Univ. Recife 45: 7. 1956. Type: *Piedraia* Fonseca & Leão

Pileolariaceae (Arthur) Cummins & Y. Hirats., Ill. Gen. Rust Fungi, ed. 2: 14. 1983. Basionym: trib. *Pileolarieae* Arthur in Lotsy, Rés. Sci. Congr. Int. Bot. Vienne 1905: 340. 1906. Type: *Pileolaria* Castagne

Pilobolaceae Corda, Anleit. Stud. Mykol.: 17. Aug-Dec 1842. Type: *Pilobolus* Tode

Pilocarpaceae Zahlbr. in Engler & Prantl, Nat. Pflanzenfam. 1(1*): 116. Mar 1905. Type: *Pilocarpon* Vain.

Piptocephalidaceae J. Schröt. in Cohn, Krypt.-Fl. Schlesien 3(1): 215. 16 Aug 1886. Type: *Piptocephalis* de Bary

Placynthiaceae Å. E. Dahl in Meddel. Grønland 150(2): 49. 4 Feb 1950. Type: *Placynthium* (Ach.) Gray

Plasmodiophoraceae Zopf ex Berl. in Saccardo, Syll. Fung. 7(1): 464. 15 Mar 1888. Type: *Plasmodiophora* Woronin

Platystomaceae J. Schröt. in Cohn, Krypt.-Fl. Schlesien 3(2): 323. Nov 1894. Type: *Platystomum* Trevis.

Pleomassariaceae M. E. Barr in Mycologia 71: 949. 20 Nov 1979. Type: *Pleomassaria* Speg.

Pleosporaceae Nitschke in Verh. Naturhist. Vereines Preuss. Rheinl. Westphalens 26: 74. 1869. Type: *Pleospora* Rabenh. ex Ces. & De Not., *nom. cons.*

Pluteaceae Kotl. & Pouzar in Česká Mykol. 26: 218. 15 Oct 1972. Type: *Pluteus* Fr.

Podaxaceae Corda, Icon. Fung. 5: 24. Jun 1842. Type: *Podaxis* Desv.

Podoscyphaceae D. A. Reid in Beih. Nova Hedwigia 18: 43. 15 Mai 1965. Type: *Podoscypha* Pat.

Polyporaceae Fr. ex Corda, Icon. Fung. 3: 49. Sep-Nov 1839. Type: *Polyporus* Fr. : Fr.

Polystigmataceae Höhn. ex Nannf. in Nova Acta Regiae Soc. Sci. Upsal., ser. 4, 8: 51. Sep 1932. Type: *Polystigma* DC.

Polystomellaceae Theiss. & Syd. in Ann. Mycol. 13: 158, 160. 31 Aug 1915. Type: *Polystomella* Speg.

Porpidiaceae Hertel & Hafellner in Beih. Nova Hedwigia 79: 318. Nov 1984. Type: *Porpidia* Körb.

Protomycetaceae de Bary ex Berl. & De Toni in Saccardo, Syll. Fung. 7(1): 319. 15 Mar 1888. Type: *Protomyces* Unger

Protosteliaceae L. S. Olive in Amer. J. Bot. 49: 298. 30 Mar 1962. Type: *Protostelium* L. S. Olive & Stoian.

Protothelenellaceae Vězda & al. in Herzogia 7: 26. 15 Dec 1985. Type: *Protothelenella* Räsänen

Pseudeurotiaceae Malloch & Cain in Canad. J. Bot. 48: 1815. Oct 1970. Type: *Pseudeurotium* J. F. H. Beyma

Pseudosphaeriaceae Höhn. in Sitzungsber. Kaiserl. Akad. Wiss., Math.-Naturwiss. Cl., Abt. 1, 116: 129. 1907. Type: *Pseudosphaeria* Höhn.

Psoraceae Zahlbr. in Engler, Syllabus, ed. 2: 44. Mai 1898. Type: *Psora* Hoffm., *nom. cons.*

Pterulaceae Corner in Beih. Nova Hedwigia 33: 10. Jun 1970. Type: *Pterula* Fr. : Fr.

Pucciniaceae Chevall., Fl. Gén. Env. Paris 1: 413. 5 Aug 1826. Type: *Puccinia* Pers. : Pers.

Pucciniastraceae Gäum. ex Leppik in Ann. Bot. Fenn. 9: 139. 18 Dec 1972. Type: *Pucciniastrum* G. H. Otth

Pucciniosiraceae (Dietel) Cummins & Y. Hirats., Ill. Gen. Rust Fungi, ed. 2: 15. 1983. Basionym: trib. *Pucciniosireae* Dietel in Engler & Prantl, Nat. Pflanzenfam., ed. 2, 6: 93. 1928. Type: *Pucciniosira* Lagerh.

Pyrenophoraceae M. E. Barr in Mycologia 71: 948. 20 Nov 1979. Type: *Pyrenophora* Fr.

Pyrenotrichaceae Zahlbr. in Engler & Prantl, Nat. Pflanzenfam., ed. 2, 8: 91. Jan-Jun 1926. Type: *Pyrenothrix* Riddle

Pyrenulaceae Rabenh., Krypt.-Fl. Sachsen 2: 42. Dec 1869 - Jan 1870. Type: *Pyrenula* Ach., *nom. cons.*

Pyronemataceae Corda, Anleit. Stud. Mykol.: 149. Aug-Dec 1842. Type: *Pyronema* Carus

Pythiaceae J. Schröt. in Engler & Prantl, Nat. Pflanzenfam. 1(1): 104. Oct 1893. Type: *Pythium* Pringsh., *nom. cons.*

Pyxidiophoraceae G. R. W. Arnold in Z. Pilzk. 37: 191. Mar 1972.

Type: *Pyxidiophora* Bref. & Tavel

Radiomycetaceae Hesselt. & J. J. Ellis in Mycologia 66: 91. 27 Feb 1974. Type: *Radiomyces* Embree

Ramalinaceae C. Agardh, Aphor. Bot.: 93. 21 Dec 1821. Type: *Ramalina* Ach., *nom. cons.*

Ramariaceae Corner in Beih. Nova Hedwigia 33: 11. Jun 1970. Type: *Ramaria* Fr. ex Bonord., *nom. cons.*

Raveneliaceae Leppik in Ann. Bot. Fenn. 9: 139. 18 Dec 1972. Type: *Ravenelia* Berk.

Requienellaceae Boise in Mycologia 78: 37. 14 Feb 1986. Type: *Requienella* Fabre

Rhipidiaceae Sparrow ex Cejp in Novák, Fl. ČSR, Řada B, 2: 303, 449. 1959. Type: *Rhipidium* Cornu, *nom. cons.*

Rhizidiaceae J. Schröt. in Cohn, Krypt.-Fl. Schlesien 3(1): 189. 16 Aug 1886. Type: *Rhizidium* A. Braun

Rhizocarpaceae M. Choisy ex Hafellner in Beih. Nova Hedwigia 79: 327. Nov 1984. Type: *Rhizocarpon* Ramond ex DC.

Rhizopogonaceae Gäum. & C. W. Dodge, Compar. Morph. Fungi: 468. 1928. Type: *Rhizopogon* Fr. & Nordholm

Rhynchogastremaceae Oberw. & B. Metzler in Syst. Appl. Microbiol. 12(3): 281. 1989. Type: *Rhynchogastrema* B. Metzler & Oberw.

Rhytismataceae Chevall., Fl. Gén. Env. Paris 1: 439. 5 Aug 1826. Type: *Rhytisma* Fr. : Fr.

Richoniellaceae Jülich in Biblioth. Mycol. 85: 388. 1 Feb 1982. Type: *Richoniella* Costantin & L. M. Dufour

Rimulariaceae Hafellner in Beih. Nova Hedwigia 79: 331. Nov 1984. Type: *Rimularia* Nyl.

Roccellaceae Chevall., Fl. Gén. Env. Paris 1: 604. 5 Aug 1826. Type: *Roccella* DC., *nom. cons.*

Russulaceae Lotsy, Vortr. Bot. Stammesgesch. 1: 708. Jan-Mar 1907. Type: *Russula* Pers. : Fr.

Saccardiaceae Höhn. in Sitzungsber. Kaiserl. Akad. Wiss., Math.-Naturwiss. Cl., Abt. 1, 118: 369. 1909. Type: *Saccardia* Cooke

Saccharomycetaceae G. Winter in Rabenh. Krypt.-Fl., ed. 2, 1(1): 68. Dec 1880. Type: *Saccharomyces* Meyen ex E. C. Hansen

Saccharomycodaceae Kudrjavtsev, Syst. Hefen: 270. 1960. Type: *Saccharomycodes* E. C. Hansen

Saccharomycopsidaceae Arx & Van der Walt in Stud. Mycol. 30: 174. 31 Dec 1987. Type: *Saccharomycopsis* Schiønning

Saksenaeaceae Hesselt. & J. J. Ellis in Mycologia 66: 87. 27 Feb 1974. Type: *Saksenaea* S. B. Saksena

Saprolegniaceae Kütz. ex Warm., Haandb. Syst. Bot., ed. 2: 54. Jan-Sep 1884. Type: *Saprolegnia* Nees

Sarcoscyphaceae Le Gal ex Eckblad in Nytt Mag. Bot. 15: 103. 15 Jun 1968. Type: *Sarcoscypha* (Fr.) Boud.

Sarcosomataceae Kobayasi in J. Jap. Bot. 13: 516. Jul 1937. Type: *Sarcosoma* Casp.

Schaereriaceae M. Choisy ex Hafellner in Beih. Nova Hedwigia 79: 399. Nov 1984. Type: *Schaereria* Körb.

Schenellaceae Nann.-Bremek. in Proc. Kon. Ned. Akad. Wetensch., Sect. C, Biol. Sci., 70: 203. 1967. Type: *Schenella* T. Macbr.

Schizophyllaceae Quél., Fl. Mycol. France: 365. Apr-Jun 1888. Type: *Schizophyllum* Fr. : Fr.

Schizosaccharomycetaceae Beij. ex Klöcker in Lafar, Handb. Techn. Mykol., ed. 2, 4: 189. 1905. Type: *Schizosaccharomyces* Lindner

Schizothyriaceae Höhn. ex Trotter & al. in Saccardo, Syll. Fung. 24: 1254. 25 Apr 1928. Type: *Schizothyrium* Desm.

Sclerodermataceae Corda, Icon. Fung. 5: 23. Jun 1842. Type: *Scleroderma* Pers. : Pers.

Sclerotiniaceae Whetzel in Mycologia 37: 652. 10 Dec 1945. Type: *Sclerotinia* Fuckel, *nom. cons.*

Secotiaceae Corda, Icon. Fung. 6: 28. Oct 1854. Type: *Secotium* Kunze

Seuratiaceae Vuill. ex M. E. Barr in Mycotaxon 29: 501. 31 Jul 1987. Type: *Seuratia* Pat.

Sirobasidiaceae Lindau in Engler & Prantl, Nat. Pflanzenfam. 1(1**): 89. Dec 1897. Type: *Sirobasidium* Lagerh. & Pat.

Sirolpidiaceae Sparrow ex Cejp in Novák, Fl. ČSR, Řada B, 2: 380, 453. 1959. Type: *Sirolpidium* H. E. Petersen

Sistotremataceae Jülich in Biblioth. Mycol. 85: 390. 1 Feb 1982. Type: *Sistotrema* Fr. : Fr.

Solorinellaceae Vězda & Poelt in Phyton (Horn) 30: 48. 29 Jun 1990. Type: *Solorinella* Anzi

Sordariaceae G. Winter in Rabenh. Krypt.-Fl., ed. 2, 1(2): 162. Nov 1884. Type: *Sordaria* Ces. & De Not., *nom. cons.*

Sparassidaceae Herter in Warnstorf & al., Krypt.-Fl. Brandenburg 6: 167. 30 Nov 1910. Type: *Sparassis* Fr. : Fr.

Spathulosporaceae Kohlm. in Mycologia 65: 615. 19 Jul 1973. Type: *Spathulospora* A. R. Caval. & T. W. Johnson

Sphaerobolaceae J. Schröt. in Cohn, Krypt.-Fl. Schlesien 3(1): 688. 15 Sep 1889. Type: *Sphaerobolus* Tode : Pers.

Sphaerophoraceae Fr., Lichenogr. Eur. Reform.: 7. Jun-Jul 1831. Type: *Sphaerophorus* Pers., *nom. cons.*

Sphaerophragmiaceae Cummins & Y. Hirats., Ill. Gen. Rust Fungi, ed. 2: 15. 1983. Type: *Sphaerophragmium* Magnus

Sphinctrinaceae M. Choisy in Bull. Mens. Soc. Linn. Lyon, Soc. Bot.

19: 65. 1950. Type: *Sphinctrina* Fr. : Fr.

Spizellomycetaceae D. J. S. Barr in Canad. J. Bot. 58: 2384. 15 Nov 1980. Type: *Spizellomyces* D. J. S. Barr

Sporormiaceae Munk in Dansk Bot. Ark. 17: 450. 20 Jun 1957. Type: *Sporormia* De Not.

Steccherinaceae Parmasto, Consp. Syst. Corticiac.: 169. 29 Jul 1968. Type: *Steccherinum* Gray

Stemonitidaceae Fr., Syst. Mycol. 3: 75. 1829. Type: *Stemonitis* Roth

Stephanosporaceae Oberw. & E. Horak in Pl. Syst. Evol. 131: 162. 28 Feb 1979. Type: *Stephanospora* Pat.

Stereaceae Pilát in Hedwigia 70: 34. Mar 1930. Type: *Stereum* Pers.

Stereocaulaceae Chevall., Fl. Gén. Env. Paris 1: 596. 5 Aug 1826. Type: *Stereocaulon* Hoffm., *nom. cons.*

Stictaceae (Nyl.) Stizenb. in Ber. Thätigk. St. Gallischen Naturwiss. Ges. 1861-1862: 174. 1862. Basionym: subtrib. *Stictinae* Nyl. in Flora 43: 65. 7 Feb 1860. Type: *Sticta* (Schreb.) Ach.

Strigulaceae Zahlbr. in Engler, Syllabus, ed. 2: 46. Mai 1898. Type: *Strigula* Fr. : Fr.

Strobilomycetaceae J.-E. Gilbert, Bolets: 105. 1931. Type: *Strobilomyces* Berk.

Strophariaceae Singer & A. H. Sm. in Mycologia 38: 503. 4 Oct 1946. Type: *Stropharia* (Fr.) Quél.

Syncephalastraceae Naumov ex R. K. Benj. in Aliso 4: 327. 26 Jun 1959. Type: *Syncephalastrum* J. Schröt.

Synchytriaceae J. Schröt. in Engler & Prantl, Nat. Pflanzenfam. 1(1): 71. Aug 1892. Type: *Synchytrium* de Bary & Woronin

Taphrinaceae Gäum. in Gäumann & Dodge, Compar. Morph. Fungi: 161. 1928. Type: *Taphrina* Fr. : Fr.

Teloschistaceae Zahlbr. in Engler, Syllabus, ed. 2: 45. Mai 1898. Type: *Teloschistes* Norman

Terfeziaceae E. Fisch. in Engler & Prantl, Nat. Pflanzenfam. 1(1): 312. Feb 1897. Type: *Terfezia* (Tul. & C. Tul.) Tul. & C. Tul.

Testudinaceae Arx in Persoonia 6: 366. 27 Dec 1971. Type: *Testudina* Bizz.

Tetragoniomycetaceae Oberw. & Bandoni in Canad. J. Bot. 59: 1039. 26 Jun 1981. Type: *Tetragoniomyces* Oberw. & Bandoni

Thamnidiaceae Fitzp., Lower Fung.: 242, 253. Jul-Dec 1930. Type: *Thamnidium* Link

Thelebolaceae (Brumm.) Eckblad in Nytt Mag. Bot. 15: 22. 15 Jun 1968. Basionym: subfam. *Theleboloideae* Brumm. in Persoonia, Suppl. 1: 59. 24 Aug 1967. Type: *Thelebolus* Tode

Thelenellaceae O. E. Erikss. ex H. Mayrhofer in Biblioth. Lichenol.

26: 16. Jan-Jul 1987. Type: *Thelenella* Nyl.

Thelephoraceae Chevall., Fl. Gén. Env. Paris 1: 84. 5 Aug 1826. Type: *Thelephora* Ehrh. ex Willd. : Fr.

Thelotremataceae (Nyl.) Stizenb. in Ber. Thätigk. St. Gallischen Naturwiss. Ges. 1861-1862: 167. 1862. Basionym: subtrib. *Thelotrematinae* Nyl., Lich. Scand.: 183. Jun-Jul 1861. Type: *Thelotrema* Ach.

Thraustochytriaceae Sparrow ex Cejp in Novák, Fl. ČSR, Řada B, 2: 77, 419. 1959. Type: *Thraustochytrium* Sparrow

Thrombiaceae Poelt & Vězda ex J. C. David & D. Hawksw. in Syst. Ascomycetum 10: 16. Jun 1991. Type: *Thrombium* Wallr.

Thyridiaceae J.-Z. Yue & O. E. Erikss. in Syst. Ascomycetum 6: 233. Dec 1987. Type: *Thyridium* Nitschke

Tilletiaceae J. Schröt. in Cohn, Krypt.-Fl. Schlesien 3(1): 276. 27 Aug 1887. Type: *Tilletia* Tul. & C. Tul.

Tilletiariaceae R. T. Moore in Bot. Mar. 23: 364. Jun 1980. Type: *Tilletiaria* Bandoni & B. N. Johri

Torrendiaceae Jülich in Biblioth. Mycol. 85: 392. 1 Feb 1982. Type: *Torrendia* Bres.

Trapeliaceae M. Choisy ex Hertel in Vortr. Gesamtgeb. Bot., ser. 2, 4: 181. 1970. Type: *Trapelia* M. Choisy

Tremellaceae Fr., Syst. Mycol. 1: lv. 1 Jan 1821. Type: *Tremella* Pers. : Fr., *nom. cons.*

Trichiaceae Chevall., Fl. Gén. Env. Paris 1: 322. 5 Aug 1826. Type: *Trichia* Haller

Trichocomaceae E. Fisch. in Engler & Prantl, Nat. Pflanzenfam. 1(1): 310. Feb 1897. Type: *Trichocoma* Jungh.

Tricholomataceae R. Heim ex Pouzar in Česká Mykol. 37: 175. 28 Jul 1983, *nom. cons.* (conserved against *Hydnangiaceae, Hygrophoropsidaceae, Lentinaceae* and *Physalacriaceae*). Type: *Tricholoma* (Fr.) Staude, *nom. cons.*

Trichopeltidaceae Theiss. in Centralbl. Bakteriol., 2. Abth. 39: 629. 6 Jan 1914. Type: *Trichopeltis* Speg.

Trichosphaeriaceae G. Winter in Rabenh. Krypt.-Fl., ed. 2, 1(2): 191. Nov 1884. Type: *Trichosphaeria* Fuckel

Trichotheliaceae (Müll. Arg.) Bitter & F. Schill. in Hedwigia 67: 273. Dec 1927. Basionym: subtrib. *Trichotheliinae* Müll. Arg. in Bot. Jahrb. Syst. 6: 418. 1885. Type: *Trichothelium* Müll. Arg.

Tripterosporaceae Cain in Canad. J. Bot. 34: 699. 2 Aug 1956. Type: *Tripterospora* Cain

Tryblidiaceae Rehm in Rabenh. Krypt.-Fl., ed. 2, 1(3): 191. Aug 1888. Type: *Tryblidium* Wallr.

Trypetheliaceae Zenker in Goebel, Pharm. Waarenk. 1: 123. 1827-1829. Type: *Trypethelium* Spreng., *nom. cons.*

Tuberaceae Dumort., Comment. Bot.: 69, 79. Nov-Dec 1822. Type: *Tuber* F. H. Wigg. : Fr.

Tubeufiaceae M. E. Barr in Mycologia 71: 948. 20 Nov 1979. Type: *Tubeufia* Penz. & Sacc.

Tubulicrinaceae Jülich in Biblioth. Mycol. 85: 392. 1 Feb 1982. Type: *Tubulicrinis* Donk

Tulasnellaceae Juel in Bih. Kongl. Svenska Vetensk.-Akad. Handl. 23(3,12): 21. 13 Oct 1897. Type: *Tulasnella* J. Schröt.

Tulostomataceae E. Fisch. in Engler & Prantl, Nat. Pflanzenfam. 1(1**): 342. Mar 1900. Type: *Tulostoma* Pers. : Pers.

Typhulaceae Jülich in Biblioth. Mycol. 85: 393. 1 Feb 1981. Type: *Typhula* (Pers.) Fr. : Fr.

Umbilicariaceae Chevall., Fl. Gén. Env. Paris 1: 640. 5 Aug 1826. Type: *Umbilicaria* Hoffm.

Uropyxidaceae (Arthur) Cummins & Y. Hirats., Ill. Gen. Rust Fungi, ed. 2: 14. 1983. Basionym: subfam. *Uropyxidoideae* Arthur in Lotsy, Rés. Sci. Congr. Int. Bot. Vienne 1905: 341. 1906. Type: *Uropyxis* J. Schröt.

Ustilaginaceae G. Winter in Rabenh. Krypt.-Fl., ed. 2, 1(1): 80. Dec 1880. Type: *Ustilago* (Pers.) Roussel

Valsaceae Tul. & C. Tul., Select. Fung. Carpol. 1: 180. 1861. Type: *Valsa* Fr., *nom. cons.*

Venturiaceae E. Müll. & Arx ex M. E. Barr in Mycologia 71: 947. 20 Nov 1979. Type: *Venturia* Sacc., *nom. cons.*

Verrucariaceae Zenker in Goebel, Pharm. Waarenk. 1: 123. 1827-1829. Type: *Verrucaria* Schrad. : Fr., *nom. cons.*

Vezdaeaceae Poelt & Vězda ex J. C. David & D. Hawksw. in Syst. Ascomycetum 10: 16. Jun 1991. Type: *Vezdaea* Tscherm.-Woess & Poelt

Vibrisseaceae Korf in Mycosystema 3: 23. 31 Dec 1990. Type: *Vibrissea* Fr. : Fr.

Vizellaceae H. J. Swart in Trans. Brit. Mycol. Soc. 57: 456. 24 Dec 1971. Type: *Vizella* Sacc.

Xenasmataceae Oberw. in Sydowia 19: 25. 15 Nov 1966. Type: *Xenasma* Donk

Xerocomaceae (Singer) Pegler & T. W. K. Young in Trans. Brit. Mycol. Soc. 76: 112. 20 Feb 1981. Basionym: subfam. *Xerocomoideae* Singer in Farlowia 2: 279. 17 Sep 1945. Type: *Xerocomus* Quél., *nom. cons.*

Xylariaceae Tul. & C. Tul., Select. Fung. Carpol. 2: 3. 1863. Type: *Xylaria* Hill ex Schrank, *nom. cons.*

Zoopagaceae Drechsler in Mycologia 30: 154. 1 Apr 1938. Type: *Zoopage* Drechsler

Zopfiaceae G. Arnaud ex D. Hawksw. in Syst. Ascomycetum 11: 77. Jun 1992. Type: *Zopfia* Rabenh.

TABLE OF CONTENTS

Addresses of the authors:

Marshall J. Crosby, Missouri Botanical Garden, P.O. Box 299, St. Louis, MO 63166, U.S.A.

John C. David, International Mycological Institute, Bakeham Lane, Egham, Surrey TW20 9TY, U.K.

Werner Greuter, Botanischer Garten und Botanisches Museum Berlin-Dahlem, Königin-Luise-Str. 6-8, W-1000 Berlin 33, Germany.

Riclef Grolle, Kernbergstr. 59 (511-48), O-6900 Jena, Germany.

Ruurd D. Hoogland, Laboratoire de Phanérogamie, Muséum national d'Histoire naturelle, 16 rue Buffon, F-75005 Paris, France.

James L. Reveal, Department of Botany, University of Maryland, College Park, MD 20742-5815, U.S.A.

Gea Zijlstra, Herbarium, Heidelberglaan 2, NL-3584 CS Utrecht, The Netherlands.